Estimator's General Construction Man-Hour Manual

Second Edition

John S. Page

G|P
P|❦ **Gulf Professional Publishing**
An Imprint of Elsevier

To my wife, Celesta, who has spent
many lonely nights that this book
might become a reality.

**Estimator's General Construction
Man-Hour Manual**
Second Edition

Permissions may be sought directly from Elsevier's Science and Technology Rights Department in
Oxford, UK. Phone: (44) 1865 843830, Fax: (44) 1865 853333, e-mail: permissions@elsevier.co.uk.
You may also complete your request on-line via the Elsevier homepage: http://www.elsevier.com by
selecting "Customer Support" and then "Obtaining Permissions".

Originally published by Gulf **Professional** Publishing,
Houston, TX.
ISBN-13: 978-0-87201-320-9
ISBN-10: 0-87201-320-0
Transferred to Digital Printing, 2010
Printed and bound in the United Kingdom
For information, please contact:
Manager of Special Sales
Butterworth–Heinemann
225 Wildwood Avenue
Woburn, MA 01801–2041
Tel: 781-904-2500
Fax: 781-904-2620
For information on all Butterworth–Heinemann publications
available, contact our World Wide Web home page at:
http://www.bh.com

This book was
reviewed by the author
and reprinted
August 1997.

CONTENTS

Section 1 – DEMOLITION

Section 2 – SITE-GRADING & STRUCTURAL EXCAVATION

Section 3 – SITE DRAINAGE PIPE & STRUCTURES

Section 4 – SITE IMPROVEMENTS

Section 5 – SHEET & FOUNDATION PILING

Section 6 – FORMWORK

Section 7 – REINFORCING STEEL AND MESH

Section 8 – MISCELLANEOUS EMBEDDED ITEMS

Section 9 – CONCRETE

Section 10 – MASONRY

Section 11 – STRUCTURAL STEEL & MISCELLANEOUS IRON

Section 12 - CARPENTRY & MILLWORK

Section 13 - METAL DOORS, SASH, GLASS, & GLAZING

Section 14 - SPECIAL FLOORING

Section 15 – SPECIAL WALLS & CEILINGS

Section 16 – ROOFING & SIDING

Section 17 – ORNAMENTAL METAL & SPECIAL PARTITION

Section 18 – PAINTING

PREFACE

It is not the intention of this manual to hold anything new for the top flight general construction estimator whose ability, know-how and knowledge in the Industry is the product of many years of schooling, actual competitive bidding, hard knocks and time-consuming analyses of both good and bad estimates. This type of estimator knows that to prepare an accurate labor estimate in dollar value one must first have a basis or reason for the use of monetary units.

Simply to say that a unit or block of work is worth so many dollars because it cost your company that on a previous project is absurd, ridiculous and tends to show the weakness of the inexperienced estimator. The purpose of this manual is to offer assistance or a basis, in direct labor manhours, for this type estimator.

The following direct labor manhours, or in some cases, comparison percentages, are the product of obtaining many hundreds of time and methods and preplanned studies, coupled with actual cost of various operations both in the field and fabricating shops, on many varied types of large commercial and industrial projects throughout the country, ranging in cost from $10,000 to $100,000,000.

After careful analysis of these reports we found that a productivity of seventy (70) percent was a fair average for all crafts with the exception of brick masons whose productivity was equal to only fifty-five (55) percent. The direct labor manhour tables throughout this manual are based on these percentages.

We do not attempt to define construction items or supply description or directions as to how a take-off should be made, or what quantity or quality of material make up a unit or block of work or how a particular unit or block of work should be constructed. Before anyone engaged in construction work becomes an estimator, even inexperienced, he must have some knowledge of this.

Neither will you find any cost as to materials, equipment usage, warehousing, and storage, fabricating shop set-up or overhead. If a material take-off is available this cost can be obtained, at current prices, from vendors who are to furnish the materials. Warehousing and storage, fabricating shop set-up, equipment usage and overhead can readily be obtained by a good estimator who can visualize and consider the job schedule size and location. These are items which can and must be considered for the individual project.

Before an attempt is made to apply the following direct labor manhour tables we caution the estimator to be thoroughly familiar with the introduction on the following pages entitled "Production and Composite Rate", which is the answer to this type of estimating.

INTRODUCTION
Production and Composite Rate

Here is the switch that will turn on the light and show the way to correct application of the many manhour tables that follow.

There should be sound reasoning and understanding to back up a monetary unit before it is applied to an item for labor value. The best reasoning that we know is manhours based on what we call productivity efficiency coupled with production elements.

After comparison of many projects, constructed under varied conditions, we have found that production elements can be grouped into six different classifications and that production percentages can be classified into five different categories.

The six different classifications of production elements are:

1. GENERAL ECONOMY
2. PROJECT SUPERVISION
3. LABOR RELATIONS
4. JOB CONDITIONS
5. EQUIPMENT
6. WEATHER

The five ranges of productivity efficiency percentages are:

Type	Percentage Range
1. Very low	10 through 40
2. Low	41 through 60
3. Average	61 through 80
4. Very good	81 through 90
5. Excellent	91 through 100

Since there is such a wide range between the productivity percentages, let us attempt to evaluate each of the six elements, giving an example with each, and see just how a true productivity percentage can be obtained.

1. <u>GENERAL ECONOMY:</u> This is nothing more than the state of the nation or area in which your project is to be constructed. The things that should be reviewed and evaluated under this category are:

> a. Business trends and outlooks
> b. Construction volume
> c. The employment situation

Let us assume that after giving due consideration to these items you find them to be very good or excellent. This sounds good, but actually it means that your productivity range will be very low. This is due to the fact that with business being excellent the top supervision and craftsmen will be mostly employed and all that you will have to draw from will be inexperienced personnel. Because of this, in all probability, it will tend to create bad relationship between Owner representatives, Contract supervision, and the various craftsmen, thus making very unfavorable job conditions. On the other hand, after giving consideration to this element you may find the general economy to be of a fairly good average. Should this be the case, you should find that productivity efficiency tends to rise. This is due to the fact that under normal conditions there are enough good supervisors and craftsmen to go around, they are satisfied, thus creating good job conditions and under-standing for all concerned. We have found, in the past, that general economy of the nation or area where your project is to be constructed, sets off a chain reaction to the other five elements. We, therefore, suggest that very careful consideration be given this item.

As an example, to show how a final productivity efficiency percentage can be arrived at, let us say that we are estimating a project in a given area and after careful consideration of this element, we find it to be of a high average. Since it is of a high average, but by no means excellent, we estimate our productivity percentage at seventy-five (75) percent.

2. <u>PROJECT SUPERVISION:</u> What is the calibre of your supervision? Are they well-seasoned and experienced? What can you afford to pay them? What supply do you have to draw from? Things that should be looked at and evaluated under this element are:

> a. Experience
> b. Supply
> c. Pay

Like general economy this too must be carefully analyzed. If business is normal, in all probability, you will be able to obtain good supervision, but if business is excellent the chances are that you will have a poor lot to draw from. Should the contractor try to cut overhead by the use of cheap supervision he will usually wind up doing a very poor job. This usually results in a dissatisfied client, a loss of profit, and a loss of future work. This, like the attachment of the fee for a project, is a problem over which the estimator has no control. It must be left to Management. All the estimator can do is to evaluate and estimate his project accordingly.

To follow through with our example, after careful analysis of the three (3) items listed under this element, let us say that we have found our supervision will be normal for the project involved and we arrive at an estimated productivity rate of seventy (70) percent.

3. <u>LABOR CONDITIONS:</u> Does your organization possess a good labor relations man? Are there experienced first class satisfied craftsmen in the area where your project is to be located? Like project supervision, things that should be analyzed under this element are:

        a. Experience
        b. Supply
        c. Pay

A check in the general area where your project is to be located should be made to determine if the proper experienced craftsmen are available locally, or will you have to rely on travelers to fill your needs. Can and will your organization pay the prevailing wage rates?

For our example, let us say that for our project we have found our labor relations to be fair but feel that they could be a little better and that we will have to rely partially on travelers. Since this is the case, we arrive at an efficiency rating of sixty-five (65) percent for this element.

4. <u>JOB CONDITIONS:</u> What is the scope of your project and just what work is involved in the job? Will the schedule be tight and hard to meet, or will you have ample time to complete the project? What kind of shape or condition is the site in? Is it low and mucky and hard to drain, or is it high and dry and easy to drain? Will you be working around a plant already in production? Will there be tie-ins, making it necessary to shut down various systems of the plant? What will be the relationship between production personnel and construction personnel? Will most of your operations be manual, or mechanized? What kind of material procurement will you have? There are many items that could be considered here, dependent on the project; however, we feel that the most important items that should be analyzed under this element are as follows:

a. Scope of work
b. Site conditions
c. Material procurement
d. Manual and mechanized operations

By a site visitation and discussion with Owner representatives, coupled with careful study and analysis of the plans and specifications, you should be able to correctly estimate a productivity percentage for this item.

For our example, let us say that the project we are estimating is a completely new plant and that we have ample time to complete the project but the site location is low and muddy. Therefore, after evaluation we estimate a productivity rating of only sixty (60) percent.

5. EQUIPMENT: Do you have ample equipment to complete your project? Just what kind of shape is it in and will you have good maintenance and repair help? The main items to study under this element are:

a. Usability
b. Condition
c. Maintenance and repair

This should be the simplest of all elements to analyze. Every estimator should know what type and kind of equipment his company has, as well as what kind of mechanical shape it is in. If equipment is to be obtained on a rental basis then the estimator should know the agency he intends to use as to whether they will furnish good equipment and good maintenance.

Let us assume, for our example, that our company equipment is in very good shape, that we have an ample supply to draw from and that we have average mechanics. Since this is the case we estimate a productivity percentage of seventy (70).

6. WEATHER: Check the past weather conditions for the area in which your project is to be located. During the months that your company will be constructing, what are the weather predictions based on these past reports? Will there be much rain or snow? Will it be hot and mucky or cold and damp? The main items to check and analyze here are as follows:

a. Past weather reports
b. Rain or snow
c. Hot or cold

This is one of the worst of all elements to be considered. At best all you have is a guess. However, by giving due consideration to the items as outlined under this element, your guess will at least be based on past occurrences.

For our example, let us assume that the weather is about half good and half bad during the period that our project is to be constructed. We must then assume a productivity range of fifty (50) percent for this element.

We have now considered and analyzed all six elements and in the examples for each individual element have arrived at a productivity efficiency percentage. Let us now group these percentages together and arrive at a total percentage:

Item	Productivity Percentage
1. General economy	75
2. Production supervision	70
3. Labor relations	65
4. Job Conditions	60
5. Equipment	70
6. Weather	50
Total	390

Since there are six elements involved, we must now divide the total percentage by the number of elements to arrive at an average percentage of productivity.

$$390 \div 6 = 65 \text{ percent average productivity efficiency}$$

At this point we caution the estimator. This example has been included as a guide to show one method that may be used to arrive at a productivity percentage. The preceding elements can and must be considered for each individual project. By so doing, coupled with the proper manhour tables that follow, a good labor value estimate can be properly executed for any place in the world, regardless of its geographical location and whether it be today or twenty years from now.

COMPOSITE RATE

Next, we must consider the composite rate. In order to correctly arrive at a total direct labor cost, using the manhours as appear in the following tables, this must be done.

Most organizations consider field personnel with a rating of superintendent or greater as a part of job overhead, and that of general foreman or lower as direct job labor cost. The direct manhours as appear on the following pages have been determined on this basis. Therefore, a composite rate should be used when converting the manhours to direct labor dollars.

Again, the estimator must consider labor conditions in the area where the project is to be located. He must ask himself how many men will he be allowed to use in a crew, can he use crews with mixed crafts, and how many crews of the various crafts will he need.

Following is an example that may be used to obtain a composite rate:

We assume that a certain project has a certain amount of form work and that We will need four (4) eight-man crews, and that only one General Foreman will be needed to head the four crews:

Rate of Craft in the given area:

General Carpenter Foreman	$10.64 per hour
Carpenter Foreman	10.14 per hour
Journeyman Carpenter	9.64 per hour
Laborer	6.91 per hour
Truck Driver	7.83 per hour

Note: General Foreman and Foreman are dead weight since they do not work with their tools, however, they must be considered and charged to the composite crew.

Crew for Composite Rate:

One General Foreman	2 hours @ $10.64 =	$ 21.28
One Foreman	8 hours @ 10.14 =	81.12
Five Carpenters	8 hours @ 9.64 =	385.60
Two Laborers	8 hours @ 6.91 =	110.56
One Truck Driver	4 hours @ 7.83 =	31.32

Total for 60 hours $629.88

$629.88 ÷ 60 = $10.498 Composite Manhour Rate for 100% Time

It is well to note at this time that, as was stated in the preface to this manual the manhours are based on an average productivity of seventy (70) percent for all Crafts with the exception of brick masons, whose average productivity is based on only fifty-five

(55) percent. Therefore. the composite rate of $10.498. as figured above. becomes equal to seventy (70) percent productivity.

Let us now assume that we have evaluated a certain project to be bid and find it to be of a low average with an overall productivity rating of only sixty-five (65) percent. This means a loss of five (5) percent of time paid per man hour. Therefore, the composite rate should have an adjustment of five (5) percent as follows:

$$\$10.498 \times 105\% = \$11.02 \text{ Composite Rate for 65\% productivity.}$$

Simply by multiplying the number of manhours estimated for a given block or item of work by the arrived at composite rate, a total estimated direct labor cost, in dollar value can be easily and accurately obtained.

It is our express desire and sincere hope that the foregoing will enable the ordinary general construction estimator to turn out a better labor estimate and assist in the elimination of much guesswork.

Section 1

DEMOLITION

It is not the intent of this section to cover every conceivable type of demolition. Instead, we have covered only items which we have encountered, on various plants or projects, which had to be removed prior to the installation of new work or additions.

The following manhour tables are an average of several projects of like demolition and include only major structural items that one might have to demolish in an industrial or chemical plant.

For removal and hauling of debris, see manhour tables for hauling under Section 2 entitled "Site Grading and Structural Excavation".

1

REMOVE SITE DRAINAGE ITEMS

MANHOURS PER UNITS LISTED

Item	Unit	Manhours				
		Laborer	Air Tool Operator	Operating Engineer	Truck Driver	Total
Manholes and Catch Basins						
Remove manholes	EA	2.50	2.50	0.50	0.50	6.00
Remove catchbasins	EA	4.00	4.00	1.00	1.00	10.00
Remove frames and covers	EA	1.00	–	–	0.25	1.25
Remove V.C. or Conc. Drain Pipe						
Pipe diameter 4″	LF	0.07	–	–	0.03	0.10
Pipe diameter 6″	LF	0.07	–	–	0.03	0.10
Pipe diameter 8″	LF	0.08	–	–	0.04	0.12
Pipe diameter 10″	LF	0.09	–	–	0.04	0.13
Pipe diameter 12″	LF	0.10	–	0.10	0.10	0.30
Pipe diameter 15″	LF	0.11	–	0.11	0.11	0.33
Pipe diameter 18″	LF	0.13	–	0.13	0.13	0.39
Pipe diameter 21″	LF	0.13	–	0.13	0.13	0.39
Pipe diameter 24″	LF	0.14	–	0.14	0.14	0.42
Pipe diameter 30″	LF	0.17	–	0.17	0.17	0.51
Pipe diameter 36″	LF	0.18	–	0.18	0.18	0.54

Manhole and catch basin manhours include breaking out with air hammer and loading onto trucks for hauling.

Remove V.C. or concrete drain pipe manhours include removing and loading onto truck for hauling.

Manhours do not include truck hauling time or excavating and backfill. See respective tables for these time frames.

REMOVE MISCELLANEOUS SITE ITEMS

MANHOURS PER UNITS LISTED

Item	Unit		Manhours			
		Laborer	Air Tool Operator	Operating Engineer	Truck Driver	Total
Remove Fencing and Guard Rail						
Three strand barbed wire	LF	0.05	–	–	0.03	0.08
Five strand barbed wire	LF	0.06	–	–	0.03	0.09
Chain link fence	LF	0.10	–	–	0.05	0.15
Guard rail	LF	0.17	–	–	0.08	0.25
Remove Railroads						
Ties and track	LF	0.15	–	0.08	0.08	0.31
Turnouts	EA	16.00	–	8.00	8.00	32.00
Ballast	CY	0.05	–	0.05	0.05	0.15
Remove bituminous pavement	SY	0.08	0.04	0.04	0.08	0.24

Manhours include ample labor for the removal and loading on trucks of items as outlined above.

Manhours do not include hauling or unloading. See respective table for this item.

CORE DRILLING CONCRETE SLABS

MANHOURS REQUIRED EACH

Hole Diameter	Manhours		
	Laborer	Drill Operator	Total
Concrete Slabs to 6" Thick			
1"	1.00	1.00	2.00
2"	1.14	1.14	2.28
3"	1.33	1.33	2.66
4"	1.63	1.63	3.26
6"	2.29	2.29	4.58
8"	4.00	4.00	8.00
10"	5.33	5.33	10.66
12"	8.00	8.00	16.00
14"	8.89	8.89	17.78
16"	10.00	10.00	20.00
18"	11.43	11.43	22.86
Concrete Slabs 6" to 12" Thick			
1"	1.75	1.75	3.50
2"	2.00	2.00	4.00
3"	2.33	2.33	4.66
4"	2.85	2.85	5.70
6"	4.00	4.00	8.00
8"	7.08	7.08	14.16
10"	9.49	9.49	18.98
12"	14.24	14.24	28.48
14"	15.82	15.82	31.64
16"	17.90	17.90	35.80
18"	20.46	20.46	40.92

Manhours are average for core drilling holes in reinforced concrete slabs using methods and drills for this type of work.

Manhours do not include clean-up or loading and hauling of debris. See respective tables for these charges.

BREAK OUT CONCRETE ITEMS

MANHOURS PER UNITS LISTED

Item	Unit	Manhours					
		Laborer	Air Tool Operator	Oper. Engr.	Oiler	Iron Worker	Total
Break Out with Crane & Headache Ball							
Ground floors	cu yd	—	—	.08	.08	—	.16
Elevated slabs	cu yd	—	—	.07	.07	—	.14
Foundation piers & walls	cu yd	—	—	.20	.20	—	.40
Concrete Pavement							
With pneumatic breaker	sq yd	.13	.25	.13	—	—	.51
By hand with sledge	sq yd	2.70	—	—	—	—	2.70
Elevated Slabs							
With pneumatic breaker	cu yd	2.50	1.04	1.04	—	.50	5.08
By hand	cu yd	12.40	—	—	—	1.80	14.20
Concrete Walls							
With pneumatic breaker	cu yd	.79	.79	.79	—	—	2.37
By hand with sledge	cu yd	11.00	—	—	—	—	11.00
Remove Concrete Curbs							
With pneumatic breaker	lin ft	—	.07	.04	—	—	.11
By hand	lin ft	.45	—	—	—	—	.45
Cut Concrete Openings							
Pneumatic chisels	cu yd	—	4.05	2.03	—	—	6.08
Hand chisels	cu yd	14.00	—	—	—	—	14.00
Remove Concrete Sills	lin ft	.05	—	—	—	—	.05
Remove Concrete Loading Dock	cu yd	1.50	1.20	.85	—	.40	3.95

Manhours include breaking out of concrete items using methods as outlined above.

Manhours do not include loading and hauling of debris. See respective table for this charge.

SAW CUTTING ASPHALT, CONCRETE AND MASONRY

MANHOURS REQUIRED PER LINEAR FOOT

Item	Manhours		
	Laborer	Saw Operator	Total
Asphalt pavement—one inch thick	.036	.018	.054
Asphalt—add for each additional inch	.016	.008	.024
Conc. slab with mesh—per inch of depth	.060	.030	.090
Conc. slab with rod—per inch of depth	.148	.074	.222
Conc. walls—per inch of depth	.220	.110	.330
Conc. walls—reinforced—per inch of depth	.296	.148	.444
Brick walls—per inch of depth	.220	.110	.330
Conc. block walls—per inch of depth	.184	.092	.276

TORCH CUTTING

MANHOURS PER UNITS LISTED

Item	Unit	Ironworker Manhours
Steel plate—½ inch thick or less	LF	.500
Steel plate—¾ inch thick	LF	.550
Steel plate—1 inch thick	LF	.600
Steel plate—1¼ inch thick	LF	.700
Steel plate—1½ inch thick	LF	.750
Steel plate—1¾ inch thick	LF	.800
Steel plate—2 inches thick	LF	.900
Steel bar—½ inch round or less	EA	.034
Steel bar—⅝ to 1 inch round	EA	.050
Steel bar—1 inch square	EA	.067
Steel bar—1¼ inch square	EA	.067

Manhours include layout and cutting time for the items and quantities listed.

Manhours do not include clean-up or loading and hauling of debris. See respective tables for these charges.

MASONRY & WOOD ITEMS

MANHOURS PER UNITS LISTED

Item	Unit	Manhours		
		Laborer	Air Tool Operator	Total
Solid Masonry Walls				
Pneumatic pick	cu yd	.23	.23	.46
By hand	cu yd	1.35	–	1.35
Hollow Masonry Walls				
Pneumatic pick	cu yd	.18	.18	.36
By hand	cu yd	1.31	–	1.31
Cut Masonry Openings	cu yd	5.00	–	5.00
Remove Wood Decking	100 sq ft	.67	–	.67
Remove Roof Lumber	Mfbm*	12.00	–	12.00
Remove Wood Siding	100 sq ft	2.00	–	2.00
Remove Structural Framing	Mfbm*	8.00	–	8.00
Remove Stud Framing	Mfbm*	6.00	–	6.00

Manhours are average for breaking our or wrecking of items as listed above using methods and tools as are necessary for this type of work.

Manhours do not include cleaning of materials or loading and hauling. See respective tables for these charges.

*1,000 foot board measure

REMOVAL OF STEEL, EQUIPMENT, AND PIPE

MANHOURS REQUIRED PER UNITS LISTED

| Item | Unit | Manhours | | | | | |
		Laborer	Pipe-Fitter	Iron-Worker	Operating Engineer	Truck Driver	Total
Steel Items							
Structural	Ton	3.09	—	6.18	1.03	1.03	11.33
Miscellaneous	Ton	5.41	—	10.82	1.80	1.80	19.83
Mechanical Equipment							
Light	Ton	4.00	4.00	8.00	2.00	2.00	20.00
Heavy	Ton	3.32	3.32	6.64	1.57	1.57	16.42
Carbon Steel Pipe							
2" and below	LF	0.052	0.104	—	0.052	0.052	0.260
3" to 6"	LF	0.094	0.188	—	0.094	0.094	0.470
8" to 12"	LF	0.200	0.400	—	0.200	0.200	1.000
12" to 16"	LF	0.348	0.696	—	0.348	0.348	1.736
18" to 24"	LF	0.500	1.000	—	0.500	0.500	2.500
24" to 36"	LF	0.600	1.200	—	0.600	0.600	3.000

Manhours include removal and loadout of items as outlined.

Manhours do not include hauling time. See respective table for this charge.

REMOVE SPECIAL FLOORING

MANHOURS PER HUNDRED (100) SQUARE FEET

Item	Manhours			
	Laborer	Air Tool Operator	Oper. Engr.	Total
Remove Finish Wood Flooring	2.03	—	—	2.03
Remove Wood Block Flooring	5.50	—	—	5.50
Remove Terrazzo Floor	.69	1.39	.69	2.77
Remove Asphalt Tile	.75	—	—	.75
Remove Rubber Tile	.75	—	—	.75
Remove Brick Floor	.90	1.80	.90	3.60
Remove Ceramic Tile	.77	1.53	.77	3.07
Remove Cork Tile	.80	—	—	.80
Remove Quarry Tile	.81	1.62	.81	3.24
Remove Linoleum	.15	—	—	.15

Manhours are average for the removal of flooring and tile work as outlined above using methods and tools as may be necessary for this type of wrecking.

Manhours do not include loading or hauling. See respective table for this charge.

MISCELLANEOUS MATERIALS

MANHOURS PER HUNDRED (100) SQUARE FEET

Item	Manhours			
	Laborer	Air Tool Operator	Oper. Engr.	Total
Remove Wall Board				
Fibre board	1.03	—	—	1.03
Plaster board	1.07	—	—	1.07
Plywood	1.14	—	—	1.14
Remove Built-up Roofing	.75	—	—	.75
Remove Roof Plank				
Gypsum	.80	—	—	.80
Concrete	.90	—	—	.90
Remove Corrigated Metal				
Roofing	1.33	—	—	1.33
Siding	1.20	—	—	1.20
Remove Plaster from Masonry	1.05	—	—	1.05
Remove Paint				
Sandblast	.50	1.00	.50	2.00
By hand	3.75	—	—	3.75
Sandblast Exterior of Building	.67	1.33	—	2.00

Manhours are average for the removal of above items using necessary methods and tools as required.

Manhours do not include loading and hauling. See respective tables for this charge.

Section 2

SITE-GRADING & STRUCTURAL EXCAVATION

This section is installed for the express purpose of covering the removal, and replacement, as may be necessary, of earthwork and the removal of trees and brush.

The manhour tables that follow are an average of many projects of varied nature and include operational time for site excavation and grading as well as structural foundation excavation, backfill and grading.

No emphasis has been placed on the degree of swing or depth of cut as may pertain to the efficiency rating of a piece of heavy equipment such as a power shovel, dragline, etc. As is stated above, the manhours are average for this type of work. Should a particular job of earthwork be encountered of a very difficult nature, the estimator should give this due consideration and adjust accordingly.

Before an estimate is made on excavation, it is well to know the kind of soil that may be encountered. For this reason we have divided soil into five groups according to the difficulty experienced in excavating it. Soils vary greatly in character and no two are exactly alike.

Group 1: LIGHT SOIL — Earth which can be shoveled easily and requires no loosening, such as sand.

Group 2: MEDIUM OR ORDINARY SOILS — Type of earth easily loosened by pick. Preliminary loosening is not required when power excavating equipment such as shovels, dragline scrapers and backhoes are used. This earth is usually classified as ordinary soil and loam.

Group 3: HEAVY OR HARD SOIL — This type of soil can be loosened by pick but this loosening is sometimes very hard to do. It may be excavated by sturdy power shovels without preliminary loosening. Hard and compack loam containing gravel, small stones and boulders, stiff clay or compacted gravel are good examples of this type.

Group 4: HARD PAN OR SHALE — A soil that has hardened and is very difficult to loosen with picks. Light blasting is often required when excavating with power equipment.

11

<u>Group 5:</u> ROCK – Requires blasting before removal and transporting.   May be divided into different grades such as hard, soft or medium.

From the following manhour tables, a complete direct labor estimate can be made for excavating almost any type of soil which may be encountered.

REMOVING TREES

NET MANHOURS—EACH

Average Tree Diameter in Inches	Softwood Trees		Hardwood Trees	
	Open Area	Congested Area	Open Area	Congested Area
Cross-cut Saws				
4	1.49	1.86	1.88	2.35
6	2.26	2.83	2.82	3.53
8	3.24	4.02	4.00	4.96
10	4.10	5.08	5.00	6.20
12	4.98	6.18	6.00	7.44
14	6.39	7.86	7.70	9.47
16	7.39	9.09	8.80	10.82
18	8.32	10.23	9.90	12.18
20	10.58	12.91	12.60	15.37
24	12.71	15.51	15.12	18.45
30	17.09	20.85	20.10	24.52
36	20.50	25.01	24.12	29.43
Chain Saws				
4	0.37	0.46	0.47	0.59
6	0.57	0.71	0.71	0.89
8	0.81	1.00	1.00	1.24
10	1.03	1.28	1.25	1.55
12	1.25	1.55	1.50	1.86
14	1.60	1.97	1.93	2.37
16	1.85	2.28	2.20	2.71
18	2.08	2.56	2.48	3.05
20	2.65	3.23	3.15	3.84
24	3.18	3.88	3.78	4.61
30	4.28	5.22	5.03	6.14
36	5.13	6.26	6.03	7.36

Manhours include ax trimming, cutting down with cross-cut saws or chain saws and cutting into four feet lengths for the tree diameter sizes as listed above. Manhours are average for various heights of trees.

Manhours do not include hauling and piling and burning of trees, or branches, or the removal of stumps. See respective tables for these charges.

Open area manhours are for felling trees in wooded sections or fields where damage of other items or structures is not a factor. Congested area manhours are for felling trees near powerlines or other structures where danger or damage is a factor.

REMOVE TREE STUMPS

NET MANHOURS — EACH

Item	Laborer	Oper. Engr.	Powder Man	Total
Grub & Removal by Hand				
8" to 12" diameter	6.00	–	–	6.00
14" to 18" diameter	7.50	–	–	7.50
20" to 24" diameter	9.00	–	–	9.00
26" to 36" diameter	11.20	–	–	11.20
Blast & Pull with Tractor				
8" to 12" diameter	.83	.11	1.50	2.44
14" to 18" diameter	1.05	.23	2.33	3.61
20" to 24" diameter	1.50	.30	3.38	5.18
26" to 36" diameter	2.11	.42	4.76	7.29

Manhours include excavating and removing by hand or blasting and removing with cables and tractors of stumps for the base of tree diameters as listed above.

Manhours do not include burning or removal from premises. See respective table for burning.

CLEARING TREES & BRUSH

MANHOURS PER ACRE

Item	Manhours		
	Laborer	Oper. Engr.	Total
Clear with Dozer, Pile & Burn			
Heavy timbered	27.00	11.25	38.25
Medium timbered	16.20	6.75	22.95
Light timbered	12.15	5.07	17.22
Clear Light Trees & Brush with:			
Tractors & Cables	.53	1.05	1.58

Clear with Dozer, Pile and Burn units include:

Dozing into Piles and Burning.

Clear Light Trees and Brush units include:

Removal with tractor and cable of saplings or light trees and brush, piling and burning.

Manhours do not include sawing, cutting or trimming of trees or the removal of stumps. See respective tables for these charges.

Overtime allowance for fire tender has not been considered above. If this is necessary, add for this operation.

GENERAL SITE GRADING

MANHOURS PER HUNDRED (100) CUBIC YARDS

Item	Manhours Operating Engineer
Self-Loading Crawler Tractor, Hauled Scraper	
9 Cubic Yard Payload	
Haul 400 feet	1.10
Haul 800 feet	1.30
Haul 1000 feet	1.50
Haul 1500 feet	1.90
Haul 2500 feet	2.50
Haul 3000 feet	3.30
Haul 4000 feet	3.75
Pusher Tractor, Loaded Crawler Tractor, Hauled Scraper	
9 Cubic Yard Payload	
2 Scrapers Haul 400 feet	1.05
2 Scrapers Haul 800 feet	1.40
3 Scrapers Haul 1000 feet	1.50
4 Scrapers Haul 1500 feet	1.75
5 Scrapers Haul 2500 feet	2.00
7 Scrapers Haul 4000 feet	3.00

Manhours include necessary labor for operating and maintaining equipment for the above type of work.

Manhours are based on loading areas and haul lanes being fairly level and in average condition and soil being ordinary dry earth.

Manhours do not include spotting or grade checking.

For heavy or hard soil increase manhours eight (8) percent.

Manhours are based on equipment working at seventy-five (75) percent efficiency.

GENERAL SITE GRADING

MANHOURS PER HUNDRED (100) CUBIC YARDS

Item	Manhours Operating Engineer
Grading with Self-Loading, Self-Propelled Scraper	
<u>9 Cubic Yard Payload</u>	
Haul 400 feet	.85
Haul 800 feet	1.05
Haul 1000 feet	1.10
Haul 1500 feet	1.35
Haul 2500 feet	1.95
Haul 3000 feet	2.20
Haul 4000 feet	2.50
Grading with Pusher Tractor, Loaded Self-Propelled Scraper	
<u>9 Cubic Yard Payload</u>	
2 Scrapers Haul 800 feet	.90
2 Scrapers Haul 1000 feet	1.00
3 Scrapers Haul 1500 feet	1.10
4 Scrapers Haul 2500 feet	1.30
5 Scrapers Haul 4000 feet	1.90

Manhours include necessary labor for operating and maintaining equipment for the above type of work.

Manhours are based on loading area and haul lanes being fairly level and in average condition and soil being ordinary dry earth.

Manhours do not include spotting or grade checking.

For heavy or hard soil increase manhours eight (8) percent.

Manhours are based on equipment working at seventy-five (75) percent efficiency.

GENERAL SITE GRADING

MANHOURS PER HUNDRED (100) CUBIC YARDS

Item	Manhours Operating Engineer
Grading with Self-Loading. Self-Propelled Scraper	
11 Cubic Yard Payload	
Haul 400 feet	.73
Haul 800 feet	.90
Haul 1000 feet	.95
Haul 1500 feet	1.16
Haul 2500 feet	1.68
Haul 3000 feet	1.89
Haul 4000 feet	2.15
Grading with Pusher Tractor. Loaded Self-Propelled Scraper	
11 Cubic Yard Payload	
2 Scrapers Haul 800 feet	.77
2 Scrapers Haul 1000 feet	.86
3 Scrapers Haul 1500 feet	.95
4 Scrapers Haul 2500 feet	1.12
5 Scrapers Haul 4000 feet	1.63

Manhours include necessary labor for operating and maintaining equipment for the above type of work and are based on equipment working at seventy-five (75) percent efficiency.

Manhours are based on loading area and haul lanes being fairly level and in average condition and soil being ordinary dry earth.

Manhours do not include spotting or grade checking.

For heavy or hard soil increase above manhours ten (10) percent.

GENERAL SITE GRADING

MANHOURS PER HUNDRED (100) CUBIC YARDS

Item	Manhours Operating Engineer
Self-Loading Crawler Tractor. Hauled Scraper	
11 Cubic Yard Payload	
Haul 400 feet	.95
Haul 800 feet	1.12
Haul 1000 feet	1.29
Haul 1500 feet	1.63
Haul 2500 feet	2.15
Haul 3000 feet	2.84
Haul 4000 feet	3.23
Pusher Tractor. Loaded Crawler Tractor. Hauled Scraper	
11 Cubic Yard Payload	
2 Scrapers Haul 400 feet	.90
2 Scrapers Haul 800 feet	1.20
3 Scrapers Haul 1000 feet	1.29
4 Scrapers Haul 1500 feet	1.51
5 Scrapers Haul 2500 feet	1.72
7 Scrapers Haul 4000 feet	2.58

Manhours include necessary labor for operating and maintaining equipment for the above type of work and are based on equipment working at seventy-five (75) per cent efficiency.

Manhours are based on loading areas and haul lanes being fairly level and in average condition and soil being ordinary dry earth.

Manhours do not include spotting or grade checking.

For heavy or hard soil increase above manhours ten (10) percent.

GENERAL SITE GRADING

MANHOURS PER HUNDRED (100) CUBIC YARDS

Item	Manhours Operating Engineer
Grading with Self-Loading. Self-Propelled Scraper	
23 Cubic Yard Payload	
Haul 400 feet	.43
Haul 800 feet	.53
Haul 1000 feet	.55
Haul 1500 feet	.68
Haul 2500 feet	.98
Haul 3000 feet	1.10
Haul 4000 feet	1.25
Grading with Pusher Tractor. Loaded Self-Propelled Scraper	
23 Cubic Yard Payload	
2 Scrapers Haul 800 feet	.45
2 Scrapers Haul 1000 feet	.50
3 Scrapers Haul 1500 feet	.55
4 Scrapers Haul 2500 feet	.65
5 Scrapers Haul 4000 feet	.95

Manhours include necessary labor for operating and maintaining equipment for the above type of work and are based on equipment working at seventy-five (75) percent efficiency.

Manhours are based on loading area and haul lanes being fairly level and in average condition and soil being ordinary dry earth.

Manhours do not include spotting or grade checking.

For heavy or hard soil increase above manhours fifteen (15) percent.

GENERAL SITE GRADING

MANHOURS PER HUNDRED (100) CUBIC YARDS

Item	Manhours Operating Engineer
Self-Loading Crawler Tractor. Hauled Scraper	
23 Cubic Yard Payload	
Haul 400 feet	.55
Haul 800 feet	.65
Haul 1000 feet	.75
Haul 1500 feet	.95
Haul 2500 feet	1.25
Haul 3000 feet	1.65
Haul 4000 feet	1.88
Pusher Tractor. Loaded Crawler Tractor. Hauled Scraper	
23 Cubic Yard Payload	
2 Scrapers Haul 400 feet	.53
2 Scrapers Haul 800 feet	.70
3 Scrapers Haul 1000 feet	.75
4 Scrapers Haul 1500 feet	.88
5 Scrapers Haul 2500 feet	1.00
7 Scrapers Haul 4000 feet	1.50

Manhours include necessary labor for operating and maintaining equipment for the above type of work and are based on equipment working at seventy-five (75) percent efficiency.

Manhours are based on loading areas and haul lanes being fairly level and in average condition and soil being ordinary dry earth.

Manhours do not include spotting or grade checking.

For heavy or hard soil increase above manhours fifteen (15) percent.

SHAPE, COMPACT & FINE GRADE FILL & REMOVE TOP SOIL

MANHOURS PER UNITS LISTED

Item	Unit	Manhours			
		Oper. Engr.	Laborer	Truck Driver	Total
Shape and Compact Fill					
Motor grader	100 sq yds	.50	—	—	.50
Sheeps foot roller	100 sq yds	.45	—	—	.45
Drum roller	100 sq yds	.40	—	—	.40
Sprinkling	100 sq yds	—	—	.20	.20
Fine Grade					
By hand	100 sq ft	—	.75	—	.75
Motor grader or bulldozer	100 sq yds	1.30	—	—	1.30
Remove Top Soil and Grade	100 sq ft	2.30	—	—	2.30

Manhours include labor as may be necessary for the above described operations.

Manhours do not include excavation or backfill. See respective tables for these charges.

Remove Top Soil and Grade manhours includes the removal of up to six (6) inches of top soil.

MAJOR FILLS & STRIPPINGS FOR BUILDING SITES

MANHOURS PER HUNDRED (100) SQUARE YARDS

Item	Manhours
	Operating Engineer
Tractor and Scraper (70-80 hp)	
Haul 400 feet	.20
Haul 800 feet	.25
Haul 1000 feet	.28
Haul 1500 feet	.35
Haul 2500 feet	.50
Haul 3000 feet	.65
Haul 4000 feet	.85
Self-Propelled Scraper with Helper Tractor	
Haul 400 feet	.22
Haul 800 feet	.28
Haul 1000 feet	.30
Haul 1500 feet	.38
Haul 2500 feet	.53
Haul 3000 feet	.70
Haul 4000 feet	.90

Manhours include necessary time allotment for an average 6-inch cut in one location and hauling for the lengths as outlined above and depositing or spreading.

Manhours do not include grade checker or rolling or fine grading. See respective tables for these charges.

MACHINE EXCAVATION – POWER SHOVEL

MANHOURS PER HUNDRED (100) CUBIC YARDS

Soil	Dipper Size	Manhours			
		Oper. Engr.	Oiler	Laborer	Total
Light	1 cubic yard dipper	1.1	1.1	1.1	3.3
	¾ cubic yard dipper	1.5	1.5	1.5	4.5
	½ cubic yard dipper	2.0	2.0	2.0	6.0
Medium	1 cubic yard dipper	2.0	2.0	2.0	6.0
	¾ cubic yard dipper	2.8	2.8	2.8	8.4
	½ cubic yard dipper	3.7	3.7	3.7	11.1
Heavy	1 cubic yard dipper	2.7	2.7	2.7	8.1
	¾ cubic yard dipper	3.7	3.7	3.7	11.1
	½ cubic yard dipper	4.9	4.9	4.9	14.7
Hard Pan	1 cubic yard dipper	3.4	3.4	3.4	10.2
	¾ cubic yard dipper	4.6	4.6	4.6	13.8
	½ cubic yard dipper	6.1	6.1	6.1	18.3
Rock	1 cubic yard dipper	3.4	3.4	3.4	10.2
	¾ cubic yard dipper	4.6	4.6	4.6	13.8
	½ cubic yard dipper	6.1	6.1	6.1	18.3

Manhours include operations of excavating and dumping on side lines or loading into trucks.

Manhours do not include hauling or blasting. See respective tables for these charges.

Above manhours are based on excavations up to six (6) feet. If excavations are to be greater in depth than this, the estimator should consider additional methods, planning, and equipment required.

If total excavated quantity is less than one hundred (100) cubic yards increase above units by thirty (30) percent.

MACHINE EXCAVATION — BACK HOE

MANHOURS PER HUNDRED (100) CUBIC YARDS

Soil	Bucket Size	Manhours			
		Oper. Engr.	Oiler	Laborer	Total
Light	1 cubic yard bucket	1.4	1.4	1.4	4.2
	¾ cubic yard bucket	1.5	1.5	1.5	4.5
	½ cubic yard bucket	2.0	2.0	2.0	6.0
Medium	1 cubic yard bucket	2.6	2.6	2.6	7.8
	¾ cubic yard bucket	3.8	3.8	3.8	11.4
	½ cubic yard bucket	4.4	4.4	4.4	13.2
Heavy	1 cubic yard bucket	3.5	3.5	3.5	10.5
	¾ cubic yard bucket	4.0	4.0	4.0	12.0
	½ cubic yard bucket	4.9	4.9	4.9	14.7
Hard Pan	1 cubic yard bucket	4.4	4.4	4.4	13.2
	¾ cubic yard bucket	4.6	4.6	4.6	13.8
	½ cubic yard bucket	6.1	6.1	6.1	18.3
Rock	1 cubic yard bucket	4.4	4.4	4.4	13.2
	¾ cubic yard bucket	4.6	4.6	4.6	13.8
	½ cubic yard bucket	6.1	6.1	6.1	18.3

Manhours include operations of excavating and dumping on side lines or loading into trucks.

Manhours do not include hauling or blasting. See respective tables for these charges.

Above manhours are based on excavations up to six (6) feet in depth. If excavations are to be greater in depth than this the estimator should consider methods, planning and equipment required.

If total excavation quantity is less than one hundred (100) cubic yards, increase above units by thirty (30) percent.

MACHINE EXCAVATION – TRENCHING MACHINE & DRAGLINE

MANHOURS PER HUNDRED (100) CUBIC YARDS

Soil	Item	Manhours			
		Oper. Engr.	Oiler	Laborer	Total
Light	DRAGLINE:				
	2 cubic yard bucket	0.7	0.7	0.7	2.1
	1 cubic yard bucket	1.1	1.1	1.1	3.3
	½ cubic yard bucket	2.0	2.0	2.0	6.0
Medium	2 cubic yard bucket	1.3	1.3	1.3	3.9
	1 cubic yard bucket	2.0	2.0	2.0	6.0
	½ cubic yard bucket	3.7	3.7	3.7	11.1
Heavy	2 cubic yard bucket	1.7	1.7	1.7	5.1
	1 cubic yard bucket	2.7	2.7	2.7	8.1
	½ cubic yard bucket	4.9	4.9	4.9	14.7
Medium	TRENCHING MACHINE:	3.8	–	7.5	11.3
Heavy	TRENCHING MACHINE:	4.8	–	9.4	14.2

Manhours include operations of excavating and dumping on side lines or loading into trucks for dragline excavation.

Manhours for trenching machine include regular trenching up to 3'-6" wide.

Manhours do not include hauling. See respective tables for this charge.

Above manhours are based on excavating up to six (6) feet deep. If excavations are to be greater in depth than this, the estimator should consider additional methods, planning, and equipment required.

If total excavated quantity is less than one hundred (100) cubic yards, increase above manhours by thirty (30) percent.

ROCK EXCAVATION

MANHOURS PER UNITS LISTED

Type Rock	Operation	Unit	Laborer	Air Tool Operator	Oper. Engr.	Powder Man	Total
	Manhours						
Soft	**Drilling Holes**						
	2½" with jackhammer	1 in ft	.06	.06	.06	—	.18
	2" with jackhammer	1 in ft	.04	.04	.04	—	.12
	2" with wagon drill	1 in ft	.01	—	.07	—	.08
	Blasting	cu yd	.04	—	—	.02	.06
Medium	**Drilling Holes**						
	2½" with jackhammer	1 in ft	.08	.08	.08	—	.24
	2" with jackhammer	1 in ft	.07	.07	.07	—	.21
	2" with wagon drill	1 in ft	.03	—	.10	—	.13
	Blasting	cu yd	.06	—	—	.02	.08
Hard	**Drilling Holes**						
	2½" with jackhammer	1 in ft	.10	.10	.10	—	.30
	2" with jackhammer	1 in ft	.09	.09	.09	—	.27
	2" with wagon drill	1 in ft	.05	—	.15	—	.20
	Blasting	cu yd	.09	—	—	.04	.13

Manhours are for the above operations for primary blasting only. If secondary blasting is required and this same method is used, increase above manhours 50 percent. If heavy weight or headache ball and crane are used for secondary breakage, refer to table under demolition for breakage of concrete slabs using this method.

Manhours do not include loading or hauling of blasted materials. See respective tables for these charges.

HAND EXCAVATION

MANHOURS PER CUBIC YARD

Soil	Excavation	Manhours		
		First Lift	Second Lift	Third Lift
Light	General Dry	1.07	1.42	1.89
	General Wet	1.60	2.13	2.83
	Special Dry	1.34	1.78	2.37
Medium	General Dry	1.60	2.19	2.83
	General Wet	2.14	2.85	3.79
	Special Dry	2.00	2.49	3.31
Hard or Heavy	General Dry	2.67	3.55	4.72
	General Wet	3.21	4.27	5.68
	Special Dry	2.94	3.91	5.70
Hard Pan	General Dry	3.74	4.97	6.61
	General Wet	4.28	5.69	7.57
	Special Dry	4.01	5.33	7.09

Manhours include picking and loosening where necessary and placing on bank out of way of excavation, or loading into trucks or wagons for hauling away.

Manhours do not include blasting, hauling or unloading. See respective tables for these charges.

DISPOSAL OF EXCAVATED MATERIALS

MANHOURS PER HUNDRED (100) CUBIC YARDS

Truck Capacity and Length of Haul	Manhours								
	Average Speed 10 mph			Average Speed 15 mph			Average Speed 20 mph		
	Truck Driver	Laborer	Total	Truck Driver	Laborer	Total	Truck Driver	Laborer	Total
3 Cu Yd Truck:									
1 Mile Haul	15.0	2.8	17.8	11.6	2.8	14.4	10.5	2.8	13.3
2 Mile Haul	21.8	2.8	24.6	16.2	2.8	19.0	14.0	2.8	16.8
3 Mile Haul	28.2	3.0	31.2	20.6	3.0	23.6	17.3	3.0	20.3
4 Mile Haul	36.0	3.0	39.0	26.8	3.0	29.8	21.0	3.0	24.0
5 Mile Haul	41.7	2.5	44.2	31.00	2.5	33.5	25.5	2.5	28.0
4 Cu Yd Truck:									
1 Mile Haul	11.3	2.1	13.4	8.8	2.0	10.8	7.9	2.1	9.0
2 Mile Haul	16.2	2.1	18.3	12.0	2.0	14.0	10.4	2.1	12.5
3 Mile Haul	21.6	2.0	23.6	15.8	2.3	18.1	13.2	2.2	15.4
4 Mile Haul	26.4	2.0	28.4	18.7	2.3	21.0	15.6	2.2	17.8
5 Mile Haul	31.3	1.3	32.6	22.2	1.6	23.8	18.5	1.5	20.0
5 Cu Yd Truck:									
1 Mile Haul	9.0	1.7	10.7	7.0	1.7	8.7	6.3	1.6	7.9
2 Mile Haul	13.0	1.7	14.7	9.7	1.7	11.4	8.3	1.7	10.0
3 Mile Haul	17.1	1.8	18.9	12.3	1.8	14.1	10.4	1.7	12.1
4 Mile Haul	21.0	2.0	23.0	15.0	2.0	17.0	12.4	1.7	14.1
5 Mile Haul	25.0	1.7	26.7	17.9	1.7	19.6	14.8	1.6	16.4
8 Cu Yd Truck:									
1 Mile Haul	5.6	1.0	6.6	4.8	1.0	5.8	4.0	1.0	5.0
2 Mile Haul	8.2	1.0	9.2	6.0	1.0	7.0	5.2	1.0	6.2
3 Mile Haul	10.5	1.1	11.6	7.8	1.1	8.9	6.5	1.0	7.5
4 Mile Haul	13.2	1.1	14.3	9.2	1.1	10.3	7.6	1.0	8.6
5 Mile Haul	15.6	1.3	16.9	10.9	1.3	12.2	9.0	1.1	10.1

Manhours include round trip for truck driver, spotting at both ends, unloading and labor for minor repairs.

Manhours do not include labor for excavation or loading of trucks. See respective tables for these charges.

DISPOSAL OF EXCAVATED MATERIALS

MANHOURS PER HUNDRED (100) CUBIC YARDS

Truck Capacity and Length of Haul	Manhours								
	Average Speed 20 mph			Average Speed 25 mph			Average Speed 30 mph		
	Truck Driver	Laborer	Total	Truck Driver	Laborer	Total	Truck Driver	Laborer	Total
3 Cu Yd Truck									
6 Mile Haul	26.0	2.5	28.5	21.4	2.5	23.9	17.6	2.4	20.0
7 Mile Haul	27.1	2.3	29.4	22.3	2.3	24.6	18.3	2.2	20.5
8 Mile Haul	28.7	2.3	31.0	23.6	2.3	25.9	19.4	2.2	21.6
9 Mile Haul	30.8	2.1	32.9	25.3	2.1	27.4	20.8	2.0	22.8
10 Mile Haul	33.4	2.1	35.5	27.5	2.1	29.6	22.6	2.0	24.6
4 Cu Yd Truck									
6 Mile Haul	19.3	1.5	20.8	16.0	2.0	18.0	13.4	1.9	15.3
7 Mile Haul	21.0	1.5	22.5	17.4	2.0	19.4	14.8	1.9	16.7
8 Mile Haul	23.5	1.5	25.0	19.5	1.8	21.3	16.6	1.8	18.4
9 Mile Haul	26.9	1.3	28.2	22.3	1.8	24.1	19.0	1.8	20.8
10 Mile Haul	31.1	1.3	32.4	25.8	1.6	27.4	21.9	1.5	23.4
5 Cu Yd Truck									
6 Mile Haul	15.6	1.4	17.0	14.1	1.6	15.7	11.7	1.5	13.2
7 Mile Haul	17.3	1.4	18.7	15.5	1.5	17.0	12.9	1.4	14.3
8 Mile Haul	19.8	1.3	21.1	17.5	1.5	19.0	14.5	1.3	15.8
9 Mile Haul	23.2	1.3	24.5	20.3	1.2	21.5	16.8	1.1	17.9
10 Mile Haul	27.4	1.2	28.6	23.7	1.2	24.9	19.7	1.1	20.8
8 Cu Yd Truck									
6 Mile Haul	9.8	1.2	11.0	9.2	1.0	10.2	7.7	1.0	8.7
7 Mile Haul	11.5	1.2	12.7	10.6	1.0	11.6	8.9	1.0	9.9
8 Mile Haul	14.1	1.1	15.2	12.7	1.0	13.7	10.6	0.9	11.5
9 Mile Haul	17.4	1.1	18.5	15.4	0.9	16.3	12.9	0.9	13.8
10 Mile Haul	21.7	1.0	22.7	19.0	0.9	19.9	15.9	0.8	16.7

Manhours include round trip for truck driver, spotting at both ends, unloading, and labor for minor repairs.

Manhours do not include labor for excavation or loading of trucks. See respective tables for these charges.

MACHINE & HAND BACKFILL

Average for Sand or Loam, Ordinary Soil, Heavy Soil and Clay

MANHOURS PER UNITS LISTED

Item	Unit	Manhours			
		Laborer	Oper. Engr.	Oiler	Total
Hand Place	cu yd	.55	—	—	.55
Bulldoze Loose Material	100 cu yds	—	3.32	—	3.32
Clamshell					
1 cubic yard bucket	100 cu yds	—	1.60	1.60	3.20
¾ cubic yard bucket	100 cu yds	—	2.00	2.00	4.00
½ cubic yard bucket	100 cu yds	—	2.75	2.75	5.50
Hand Spread					
Stone or gravel fill	cu yd	.40	—	—	.40
Sand fill	cu yd	.35	—	—	.35
Cinder fill	cu yd	.40	—	—	.40
Tamp by Hand	cu yd	.60	—	—	.60
Pneumatic Tamping	cu yd	.25	—	—	.25

Hand Place units include:, Placing by hand with shovels loose earth within hand-throwing distance of stockpiles. This unit does not include compaction.

Bulldoze Loose Material units include: The moving of pre-stockpiled loose earth over an area.

Clamshell units include: The placement of materials from reachable stockpiles.

Stone, Sand and Cinder Spread units include: The hand placing, with shovels, these materials from strategically located stockpiles.

Tamp By Hand and Pneumatic Tamping units include: The compacting of pre-spread materials in 6" layers. Manhours above for this type work are shown as laborer hours. Should air tool operator be required for this work — substitute his time for above laborer hours.

Manhours do not include trucking or fine grading. See respective tables for these charges.

LOADING DIRT FROM STOCKPILE
WITH CLAMSHELL

MANHOURS PER HUNDRED (100) CUBIC YARDS

Item	Manhours		
	Oper. Engr.	Oiler	Total
Clamshell			
2 cubic yard bucket	1.10	1.10	2.20
1 cubic yard bucket	2.00	2.00	4.00
¾ cubic yard bucket	2.50	2.50	5.00
½ cubic yard bucket	3.30	3.30	6.60

Manhours are for loading dirt from stock pile into trucks for hauling, using the type of equipment as outlined above.

Manhours do not include spotting trucks or hauling. See respective tables for these charges.

DIKE CONSTRUCTION FOR STORAGE TANKS

MANHOURS REQUIRED PER UNITS LISTED

Item	Unit	Manhours			
		Laborer	Operator Engineer	Truck Driver	Total
Grade, Fill & Compact					
Dike 2000 LF.	CY	0.126	0.050	0.025	0.201
Dike 1500 LF.	CY	0.141	0.057	0.028	0.226
Dike 1000 LF.	CY	0.157	0.063	0.031	0.251
Dike 500 LF.	CY	0.188	0.076	0.037	0.301
Dike 250 LF.	CY	0.220	0.088	0.043	0.351
Soil Poisoning					
Dike 2000 LF.	SY	0.062	–	0.002	0.064
Dike 1500 LF.	SY	0.070	–	0.003	0.073
Dike 1000 LF.	SY	0.078	–	0.003	0.081
Dike 500 LF.	SY	0.094	–	0.004	0.098
Dike 250 LF.	SY	0.109	–	0.004	0.113
Paving					
Dike 2000 LF.	SY	0.090	0.008	0.008	0.106
Dike 1500 LF.	SY	0.101	0.009	0.009	0.119
Dike 1000 LF.	SY	0.112	0.010	0.010	0.132
Dike 500 LF.	SY	0.134	0.012	0.012	0.158
Dike 250 LF.	SY	0.157	0.014	0.014	0.185

Manhours are based on a pre-cleared construction site with no obstructions and pre-compacted.

Manhours include:

Grade Fill and Compact—Cutting and moving good fill material within five or six hundred feet of the dike construction using scrapers suited for this type of work.

Soil Poisoning—Applying soil poisoning on bottom prior to placement of dike fill, on dike slopes, and on top after placement of dike fill.

Paving—Placement of two inches of sand asphalt on slopes and top of dike fill.

GUNITE BANKS FOR SLOPE PROTECTION

MANHOURS REQUIRED PER SQUARE FOOT

Item	Manhours					
	Truck Driver	Iron Worker	Laborer	Cement Finisher	Operator Engineer	Total
To 5,000 SF.						
Placing reinforcing	0.010	0.040	–	–	–	0.050
Mixing gunite	0.005	–	0.027	–	–	0.032
Placing gunite	–	–	0.028	0.042	0.014	0.084
Total	0.015	0.040	0.055	0.042	0.014	0.166
5,000 to 10,000 SF.						
Placing reinforcing	0.008	0.032	–	–	–	0.040
Mixing gunite	0.004	–	0.022	–	–	0.026
Placing gunite	–	–	0.023	0.034	0.011	0.068
Total	0.012	0.032	0.045	0.034	0.011	0.134
10,000 to 50,000 SF.						
Placing reinforcing	0.007	0.031	–	–	–	0.038
Mixing gunite	0.003	–	0.021	–	–	0.024
Placing gunite	–	–	0.022	0.033	0.010	0.065
Total	0.010	0.031	0.043	0.033	0.010	0.127
50,000 to 100,000 SF.						
Placing reinforcing	0.006	0.029	–	–	–	0.035
Mixing gunite	0.003	–	0.019	–	–	0.022
Placing gunite	–	–	0.020	0.030	0.009	0.059
Total	0.009	0.029	0.039	0.030	0.009	0.116

Manhours include handling and hauling of materials from job storage to erection site and complete installation in accordance with the following:

Placing Reinforcing—Fabrication and placement of reinforcing rod in toe, and top beam, and mesh on slope section.

Mixing Gunite—Preparing and mixing gunite for placement.

Placing Gunite—Placing three-inch thickness of air blown mortar with the use of a gunite rig and compressor.

Section 3

SITE DRAINAGE PIPE AND STRUCTURES

This section includes the installation time of various types of drainage piping and structures such as headwalls, catch basins, and manholes.

The following manhour tables are average for the work of many projects installed under varied conditions where strict methods and planning were followed along with strict reporting and recording of time and cost.

The listed manhours include time allowance to complete all necessary labor for the particular operation as may be outlined in the various tables and in accordance with the notes thereon.

CAST IRON SOIL PIPE

MANHOURS REQUIRED

Manhours Per Foot		Manhours Per Make-up			
Pipe Size (inches)	Pipe Set and Align	Lead Joint	Cement Joint	Bituminous Joint	Rubber Slip Joint
2	0.08	0.40	0.26	0.20	0.18
3	0.12	0.44	0.30	0.22	0.20
4	0.14	0.50	0.35	0.25	0.23
5	0.15	0.54	0.36	0.27	0.24
6	0.18	0.57	0.37	0.29	0.26
8	0.23	0.70	0.50	0.35	0.32
10	0.30	0.88	0.63	0.44	0.40
12	0.36	1.05	0.75	0.52	0.47
15	0.45	1.31	0.94	0.66	0.60

Manhours include handling, hauling, setting and aligning in trench, and make-up of joint as outlined above.

Manhours do not include excavation or backfill. See respective tables for these charges.

CONCRETE DRAIN PIPE

MANHOURS FOR SIZE AND UNITS LISTED

Pipe Size (inches)	Reinforced		Non-Reinforced	
	Set and Align Pipe (per foot)	Grout Joint (each)	Set and Align Pipe (per foot)	Grout Joint (each)
6	–	–	0.08	0.29
8	–	–	0.10	0.35
10	–	–	0.11	0.43
12	0.16	0.65	0.15	0.62
15	0.20	0.93	0.19	0.89
18	0.29	1.20	0.28	1.14
21	0.32	1.45	0.29	1.38
24	0.35	1.79	0.32	1.63
27	0.37	2.01	0.34	1.83
30	0.39	2.15	0.35	1.95
33	0.44	2.37	0.38	2.15
36	0.46	2.48	0.40	2.25

REINFORCED ELLIPTICAL CONCRETE PIPE

MANHOURS FOR SIZE AND UNITS LISTED

Pipe Size—Inches	Set and Align Pipe (per foot)	Grout Joint (each)
14x23 inside, round equivalent 18" dia.	0.26	1.03
24x38 inside, round equivalent 30" dia.	0.31	1.56
29x45 inside, round equivalent 36" dia.	0.37	1.80
38x60 inside, round equivalent 48" dia.	0.49	2.40
48x76 inside, round equivalent 60" dia.	0.61	3.00
58x91 inside, round equivalent 72" dia.	0.73	3.60

Manhours include handling, hauling, setting in trench, and aligning. Manhours for joint or connection of fittings are for one make-up only.

For sizes 18-inches and larger, and all elliptical pipe, allowance has been included for the placement of pipe into ditch with rig.

Manhours do not include excavation or backfill. See respective tables for these charges.

VITRIFIED CLAY AND ASBESTOS CEMENT DRAIN PIPE

MANHOURS FOR SIZE AND UNITS LISTED

Pipe Size (inches)	Vitrified Clay		Asbestos Cement	
	Set and Align Pipe (per foot)	Make-up Joint (each)	Set and Align Pipe (per foot)	Make-up Joint (each)
6	0.07	0.29	0.06	0.26
8	0.07	0.35	0.06	0.31
10	0.08	0.43	0.07	0.37
12	0.10	0.62	0.09	0.53
15	0.11	0.89	0.09	0.71
18	0.14	1.14	0.11	0.89
21	0.19	1.38	0.15	1.08
24	0.25	1.63	0.20	1.27
30	0.30	1.95	0.21	1 37
36	0.35	2.25	0.25	1.58

Manhours include handling, hauling, setting in trench, and aligning.

Manhours for joint or connection of fittings are for one make-up only.

For sizes 18-inches and larger, allowance has been included for the placement of pipe into ditch with a rig.

Manhours do not include excavation or backfill. See respective tables for these charges.

CORRUGATED METAL DRAIN PIPE

MANHOURS FOR SIZE AND UNITS LISTED

Pipe Size (inches)	Manhours			
	Set and Align Pipe (per foot)	Place Band Couplers (each)	Place Tapered End Sections (each)	Place Fittings (each)
6	0.070	0.053	–	0.210
8	0.095	0.071	–	0.285
10	0.107	0.080	–	0.321
12	0.140	0.105	2.800	0.420
15	0.177	0.155	3.500	0.620
18	0.212	0.180	4.250	0.742
21	0.245	0.215	4.950	0.858
24	0.298	0.224	5.650	0.894
30	0.352	0.264	6.30	1.056
36	0.422	0.317	7.00	1.266
48	0.525	0.381	–	1.522
60	0.635	0.445	–	1.778
72	0.770	0.539	–	2.156

Manhours include handling, hauling, setting in trench, aligning, and connecting, where required, of items as outlined.

Tapered end section manhours include placement of toe plate extensions for termination of corrugated metal pipe.

Allowance has been made in the manhours for placement of pipe into ditch with a rig when pipe size and weight warrants it.

Fitting manhours are for placement of elbows and wyes.

Manhours do not include excavation or backfill. See respective tables for these charges.

Manhours are for the installation of bituminous coated metal pipe. For plain, galvanized metal pipe deduct ten (10) percent from unit manhours.

HAND SHAPE TRENCH BOTTOM

MANHOURS PER HUNDRED (100) LINEAR FEET

Item	Manhours
	Laborers
Shape Trench for Drain Pipes	
6" to 10" Pipe	2.25
12" to 15" Pipe	5.25
18" to 21" Pipe	7.50
24" Pipe	13.00
30" Pipe	22.10
36" Pipe and Larger	28.60

Above manhours include minor hand excavations for grade and excavations for bells where required for all types of drain lines such as tile, concrete and corrugated metal in pre-excavated ditches.

Manhours are average for one hundred (100) linear feet of ditch measured on center line.

Manhours do not include excavation of ditch or placement of pipes. See respective tables for these charges.

HEADWALL, CATCH BASIN, AND MANHOLE FORMS

MANHOUR PER SQUARE FOOT

Item	MANHOURS			
	Carpenter	Laborer	Truck Driver	Total
Square Type				
Slab on Ground				
Fabricate	.04	.01	—	.05
Erect	.03	.02	.01	.06
Strip & clean	.01	.02	.01	.04
Total	.08	.05	.02	.15
Walls				
Fabricate	.04	.01	—	.05
Erect	.05	.01	.01	.07
Strip & Clean	.01	.02	.01	.04
Total	.10	.04	.02	.16
Elevated slab with				
Wood shores				
Make - up & erect	.07	.04	.01	.12
Strip & Clean	.01	.04	.01	.06
Total	.08	.08	.02	.18

Manhours are based on the fabrication and installation of one and two-inch materials for the type of formwork as outlined above.

Manhours do not include placement of reinforcement steel or pouring concrete. See respective tables for these charges.

BOX TYPE CULVERTS

MANHOURS PER UNITS LISTED

Item	Unit	Manhours				
		Carpenter	Laborer	Iron Worker	Cement Finisher	Total
Forms						
Floors	sq ft	.09	.04	—	—	.13
Walls	sq ft	.08	.02	—	—	.10
Wing walls	sq ft	.10	.03	—	—	.13
Roof slab	sq ft	.12	.05	—	—	.17
Strip & clean	sq ft	.01	.02	—	—	.03
Reinforcing Steel						
Fabricate	cwt	—	—	.58	—	.58
Place & tie	cwt	—	—	1.10	—	1.10
Concrete						
Place ready-mix	cu yd	.05	1.35	—	—	1.40
Point & patch	sq ft	—	—	—	.03	.03

Manhours are for the fabrication and placement of above outlined operations as are necessary for the construction of box type road drainage culverts.

Manhours do not include earth work or the installation of pipe. See respective tables for these charges.

HEADWALL, CATCH BASIN, AND MANHOLE CONCRETE

MANHOURS PER CUBIC YARD

Item	MANHOURS				
	Laborer	Carpenter	Oper. Engr.	Oiler	Total
Square Type					
Slab on Ground					
Direct from truck	.53	.04	—	—	.57
Chut e	.70	—	—	—	.70
Crane & Bucket	.88	—	.03	.03	.94
Walls					
Direct from truck	.55	.04	—	—	.59
Chute	1.00	—	—	—	1.00
Crane & bucket	.99	—	.04	.04	1.07
Elevated Slabs					
Ramp & buggies	1.56	—	.16	—	1.72
Crane & Bucket	1.09	—	.10	.10	1.29

Manhours do not include fabrication or erection of formwork, the placement of reinforcing steel, or the finishing of concrete. For formwork see table under this section. For reinforcing steel and finishing of concrete see respective sections in this manual.

BRICK AND BLOCK MANHOLES AND PLASTER

MANHOURS PER UNITS LISTED

Item	Unit	Mason	Plasterer	Hod Carrier	Total
Common Brick - tapered radial walls	100 ea.	1.4	—	1.2	2.6
Concrete Brick - Tapered radial walls	100 ea.	1.6	—	1.2	2.8
Concrete Block - square walls	100 sq. ft.	5.3	—	5.0	10.3
Plaster walls	Sq. yd.	—	0.1	0.9	1.0

Above manhours include mixing mortar and handling and placing masonry units.

Manhours do not include excavation or placement of concrete items not outlined above. See respective tables for these charges.

INSTALLATION OF MANHOLE FRAMES & COVERS

MANHOURS EACH

Item	Manhours
	Iron Worker
Light Duty Type	
18" Diameter	1.40
20" - 24" Diameter	2.70
Heavy Duty Type	
24" Diameter	2.50
30" - 36" Diameter	3.40

Manhours include handling, loading, hauling, unloading and complete installation of frame and cover.

Light duty type frames and covers are usually installed in lanes where they will come in contact with light or no traffic.

Heavy duty type frames and covers are usually installed in lanes which are subject to heavy traffic.

Section 4

SITE IMPROVEMENTS

It is the purpose of this section to cover site improvement items such as roads, sidewalks, fencing, and landscaping.

The following manhour tables cover labor only for the various operations as outlined in the individual tables and in accordance with the notes thereon.

These manhours are the average of many projects for this type of work on *industrial plant sites only* and are not intended to suffice for the installation of highway work or city pavement districts.

PLACE ROAD BASE MATERIALS

MANHOURS REQUIRED PER SQUARE YARD

Item Description	Manhours				
	Operator Engineer	Truck Driver	Grade Checker	Laborer	Total
Place Base Material					
Aggregate—4" thick	0.022	0.007	0.007	0.014	0.050
Aggregate—6" thick	0.032	0.011	0.011	0.022	0.076
Aggregate—8" thick	0.043	0.014	0.014	0.029	0.100
Aggregate—10" thick	0.054	0.018	0.018	0.036	0.126
Cement Treated—6" thick	0.065	0.022	0.022	0.043	0.152
Cement Treated—7" thick	0.076	0.025	0.025	0.050	0.176
Cement Treated—8" thick	0.086	0.029	0.029	0.058	0.202
Fine Grade					
By hand	–	–	–	0.068	0.068
With Motor Grader	0.117	–	–	–	0.117

Place Base Material: Manhours include dumping, spreading, compacting, and watering down. Manhours do not include hauling of base. See disposal of excavated materials, table under site grading and structural excavation section.

Fine Grade: Manhours include grading by hand or with motor grader, all road and shoulder base material.

ASPHALT CONCRETE PAVEMENT

MANHOURS REQUIRED PER SQUARE YARD

Item Description	Manhours			
	Operator Engineer	Truck Driver	Laborer	Total
Prime and Seal Base	–	0.003	0.003	0.006
Place Asphalt Topping				
2 inches thick	0.004	0.001	0.001	0.006
3 inches thick	0.005	0.002	0.009	0.013
3½ inches thick	0.007	0.003	0.013	0.023
4 inches thick	0.009	0.003	0.015	0.027

Manhours are for priming and sealing base material and placing and finishing asphalt paving in the thickness indicated.

Manhours do not include spreading and compacting of base material. See respective table for these time frames.

CURB & GUTTER WOOD FORMS

MANHOURS PER LINEAR FOOT

Item	Manhours				
	Carpenter	Cement Finisher	Cem. Fin. Helper	Truck Driver	Total
Integral Curb to 12"					
Fabricate	.01	—	—	—	.01
Erect & grade	—	.07	.01	.01	.09
Strip & clean	—	—	.03	—	.03
Total	.01	.07	.04	.01	.13
Separate Curb to 36"					
Fabricate	.02	—	—	—	.02
Erect & grade	—	.12	.02	.01	.15
Strip & clean	—	—	.04	.01	.05
Total	.02	.12	.06	.02	.22
Combined Curb & Gutter					
Fabricate	.01	—	—	—	.01
Erect & grade	—	.12	.02	.01	.15
Strip & clean	—	—	.04	—	.04
Total	.01	.12	.06	.01	.20

It is the intent of the above manhours to cover the labor involved in the outlined operations and in accordance with the introduction to this section, using one and two inch lumber all properly braced and anchored.

For separate road forms, use integral curb form manhours for pavements up to one foot thick and separate curb form manhours for pavement over one foot thick.

Truck driver time includes hauling up to one thousand feet only. If haul is to be greater than this, due consideration should be given and extra time allowed accordingly.

If forms are to be used more than once, fabrication charges should be eliminated after initial use and reuse, and Factor Manhours on page 103 applied.

CURB & GUTTER METAL FORMS

MANHOURS PER LINEAR FOOT

Item	Manhours			
	Cement Finisher	Cem. Fin. Helper	Truck Driver	Total
Integral Curb to 12"				
Erect & grade	.05	.02	.01	.08
Strip & clean	—	.02	.01	.03
Total	.05	.04	.02	.11
Separate Curb to 36"				
Erect & grade	.09	.02	.01	.12
Strip & clean	—	.03	.01	.04
Total	.09	.05	.02	.16
Combined Curb & Gutter				
Erect & grade	.10	.02	.01	.13
Strip & clean	—	.03	.01	.04
Total	.10	.05	.02	.17

It is the intent of the above manhours to cover the labor involved in the outlined operations and in accordance with the introduction to this section using prefabricated metal forms, all properly braced and anchored.

For separate road forms, use integral curb form manhours for pavements up to one foot thick and separate curb form manhours for pavement over one foot thick.

Truck driver time includes hauling up to one thousand feet only. If haul is to be greater than this, due consideration should be given and extra time allowed accordingly.

Manhours do not include screed installation. See respective table for this charge.

CURB, GUTTER & ROAD PAVEMENT

MANHOURS PER CUBIC YARD

Item	Manhours					
	Carpenter	Laborer	Cement Finisher	Oper. Engr.	Oiler	Total
Curb & Gutter						
Direct from ready-mix truck	—	—	2.5	—	—	2.5
Job mixed - direct to forms	—	—	4.0	—	—	4.0
Job mix - using wheelbarrow	—	—	5.0	—	—	5.0
Street & Road Pavement Ready-Mix						
Chute	.04	.41	—	—	—	.45
Buggies	—	.87	—	—	—	.87
Crane & bucket	—	.38	—	.08	.08	.54
Concrete Mixer with Boom Spout & Bucket						
Loading & mixing	—	.32	—	.04	.04	.40
Placing	—	.25	—	—	—	.25

Manhours for curb and gutter work include the placement and finishing of concrete as itemized above.

Manhours for street and road pavements include the loading, mixing, and placing of concrete as outlined above.

Above manhours do not include the belting, screeding or finishing of road pavements, or the fabrication of chutes. See respective tables for these charges.

FINISH CURB, GUTTER & ROAD PAVEMENT

MANHOURS PER UNITS LISTED

Item	Unit	Manhours		
		Cement Finisher	Laborer	Total
Hand Finish Roads				
Screeding	100 sq ft	1.40	.35	1.75
Brooming	100 sq ft	.50	1.00	1.50
Machine Finish Roads				
Screeding	100 sq yds	.45	—	.45
Brooming	100 sq yds	.45	—	.45
Finish Curb & Gutter	100 sq ft	2.50	2.50	5.00
Curing Slab, Curb & Gutter	100 sq yds	—	1.50	1.50

Manhours are for all necessary labor operations for the described type of finishing and curing of concrete road slabs and curb and gutter.

Manhours do not include earthwork or placement of forms or concrete.   See respective tables for these charges.

MISCELLANEOUS PAVING ITEMS

MANHOURS REQUIRED FOR UNITS LISTED

Item Description	Unit	Manhours					
		Carpenter	Laborer	Painter	Operator Engineer	Truck Driver	Total
Headers							
Redwood 1"x4"	LF	0.080	0.050	–	–	0.020	0.150
Redwood 2"x4"	LF	0.080	0.050	–	–	0.020	0.150
Redwood 2"x6"	LF	0.120	0.050	–	–	0.020	0.190
Parking Bumpers							
Precast Concrete 4' long	EA	–	0.405	–	–	0.101	0.506
Precast Concrete 6' long	EA	–	0.607	–	–	0.101	0.708
Wood 4"x4"	EA	0.067	0.334	–	–	0.101	0.502
Wood 6"x6"	EA	0.084	0.368	–	–	0.101	0.553
Painting							
Single Stall Lines	Stall	–	–	0.209	–	–	0.209
Double Stall Lines	Stall	–	–	0.293	–	–	0.293
Arrows and Signs 3' to 5'	EA	–	–	0.650	–	–	0.650

Manhours are for the fabrication. where required. and the installation of items as outlined including job handling and hauling.

Manhours do not include placement of base and pavements. See respective tables for these time frames.

BITUMINOUS AND CONCRETE WALKS

MANHOURS REQUIRED FOR UNITS LISTED

Item Description	Unit	Manhours					
		Operator Engineer	Truck Driver	Carpenter	Laborer	Cement Finisher	Total
Hand Fine Grade	SF	–	–	–	0.009	–	0.009
Bituminous Walks							
Edge Strips	LF	–	0.020	0.080	0.050	–	0.150
Asphalt 1½" thick	SY	0.004	0.002	–	0.004	–	0.010
Asphalt 2" thick	SY	0.004	0.002	–	0.006	–	0.012
Concrete Walks							
Edge Forms–2"x4"	LF	–	0.002	0.050	0.025	–	0.077
Edge Forms–2"x6'	LF	–	0.002	0.059	0.027	–	0.088
Mesh Reinforcement	SF	–	–	–	0.017	–	0.017
Concrete–4" thick		–	–	–	0.014	–	0.014
Concrete–5" thick		–	–	–	0.018	–	0.018
Concrete–6" thick		–	–	–	0.021	–	0.021
Bit. Exp. Joint ½"x4	LF	–	–	0.018	0.018	–	0.018
Bit. Exp. Joint ½"x5"	LF	–	–	0.019	0.019	–	0.019
Bit. Exp. Joint ½"x6"	LF	–	–	0.020	0.020	–	0.020
Float and Broom	SF	–	–	–	–	0.004	0.004
Cure and Protect	SF	–	–	–	0.003	–	0.003

Manhours include all operations as may be required for the fabrication and installation of the items as listed.

Manhours do not include excavation. backfill. or the placement of base materials. See respective tables for these time requirements.

CHAIN LINK FENCING

MANHOURS REQUIRED FOR UNITS LISTED

Item Description	Unit	Manhours
Fencing—No. 2 Mesh and No. 9 or No. 11 Wire		
4 feet high	LF	0.18
6 feet high	LF	0.22
8 feet high	LF	0.26
Outriggers and 3 Strands of Barbed Wire	LF	0.18
3 inch Corner or Gate Post	EA	1.63
Gates		
4 feet wide—Walk	EA	1.63
12 feet wide—Double Drive	EA	4.89
16 feet wide—Double Drive	EA	6.53
20 feet wide—Double Drive	EA	8.17
Motor Operators for Gates	EA	24.00

Fence installation manhours include installation of two-inch line post set in concrete and stringing and fastening of wire mesh in place.

For truss rods and bracing at corner post increase fencing manhours thirty (30) percent.

Outrigger and 3 strands of barbed wire manhours include installation of outriggers on line post and stringing and fastening in place 3 strands of barbed wire.

Gate manhours include installation of gates of the size shown.

Motor operator manhours include the installation of the operators but excludes electrical installation and hook-up.

All of the manhours include on the job handling and hauling.

LANDSCAPING

MANHOURS PER THOUSAND (1000) SQUARE FEET

Item	Manhours			
	Laborers	Oper. Engr.	Truck Driver	Total
Place Top Soil				
4" Deep	7.20	—	3.60	10.80
6" Deep	8.40	—	4.19	12.59
Finishing Grading				
By motor grading	.87	.87	—	1.74
By hand with rake, etc.	9.38	—	—	9.38
Place Sodding	23.70	—	5.93	29.63
Seeding and Fertilizing	2.18	—	—	2.18

Manhours are for necessary labor as may be required for the above described operations, including hauling and placing of soil and sod.

Manhours do not include the transplanting of trees or shrubs.

Section 5

SHEET & FOUNDATION PILING

This section covers the shoring and bracing of trenches and excavations with sheet piling and the placement of structural piling for foundation support.

Separate manhour tables are included for the various operations such as setting-up and removal of driving equipment and the placement of piling by several different methods, all in accordance with the notes listed on the individual tables.

SET-UP & REMOVE PILE DRIVING EQUIPMENT

NET MANHOURS — EACH

Item	Manhours				
	Oper. Engr.	Fireman	Pile Driver	Laborer	Total
Set-up Equipment	22.00	22.00	63.00	88.00	195.00
Remove Equipment	16.00	16.00	12.00	4.00	48.00

Above manhours are based on pile driver with 75-foot heads and single acting 10,000 pound steam hammer; weight of driving ram 5000 pounds completely set up for driving and dismantled after completion of driving.

Manhours do not include moving in or moving out of rig. This must be estimated individually for each project, depending upon the length and location of move.

Manhours do not include driving of piles or cut-off. See respective tables for these charges.

HAND PLACED WOOD SHEET PILING, SHORING & BRACING

MANHOURS PER HUNDRED (100) SQUARE FEET

Item	Manhours			
	Carpenter	Laborer	Truck Driver	Total
Sheet Piling - Basements & Pits				
Placing	6.75	6.75	.30	13.80
Removing	–	4.33	.30	4.63
Sheet Piling - Trenches to 8' Deep				
Placing	5.00	5.00	.30	10.30
Removing	–	2.50	.30	2.80
Sheet Piling - Trenches over 8' Deep				
Placing	5.55	5.55	.30	11.40
Removing	–	3.20	.30	3.50
Shoring or Bracing Trenches				
Placing	2.50	2.50	.20	5.20
Removing	–	1.50	.20	1.70

Manhours include necessary labor as may be involved for the hauling, handling, fabricating, placing and removing wood sheet piling, shoring and bracing as outlined above.

Manhours do not include excavation or pumping. See respective tables for these charges.

PNEUMATIC DRIVEN WOOD SHEET PILING, SHORING & BRACING

MANHOURS PER HUNDRED (100) SQUARE FEET

Item	Manhours					
	Carpenter	Laborer	Oper. Engr.	Air Tool Operator	Truck Driver	Total
Sheet Piling - Basement & Pits						
Placing	2.80	2.80	2.80	2.80	.30	11.50
Removing	–	4.00	–	–	.30	4.30
Sheet Piling - Trenches to 8' Deep						
Placing	1.80	1.80	1.80	1.80	.30	7.50
Removing	–	2.75	–	–	.30	3.05
Sheet Piling - Trenches over 8' Deep						
Placing	2.10	2.10	2.10	2.10	.30	8.70
Removing	–	3.00	–	–	.30	3.30

Manhours include necessary labor for the handling, hauling, fabricating, placing and removing wood sheet piling shored and braced in place, using a pneumatic hammer fed by compressed air.

Manhours do not include excavating or pumping. See respective tables for these charges.

WOOD & STEEL SHEET PILING DRIVEN WITH PILE DRIVER

MANHOURS PER UNITS LISTED

Item	Unit	Manhours						
		Carpenter	Laborer	Pile Driver	Oper. Engr.	Oiler	Iron Worker	Total
Wood - 20' or Deeper								
Place & drive	100-sq ft	2.00	5.00	4.00	2.00	2.00	—	15.00
Pull or remove	100-sq ft	—	2.00	1.60	.80	.80	—	5.20
Steel - 20' or Deeper								
Place & drive	100-sq ft	—	3.40	2.40	1.25	1.25	—	8.30
Pulling	100-sq ft	—	1.80	1.30	.75	.75	—	4.60
Cut-off with torch	100 lin ft	—	—	—	—	—	7.20	7.20

Manhours include handling, hauling, fabricating if necessary, placing and removing or cutting off including necessary bracing or supporting.

Manhours do not include the setting-up or dismantling of pile driver.

Manhours do not include excavating or pumping. See respective tables for these charges.

WOOD POLE & STEEL SHELL PILING

MANHOURS PER UNITS LISTED

Item	Unit	Manhours						
		Oper. Engr.	Fireman	Pile Driver	Laborer	Iron Worker	Truck Driver	Total
Wood Pole Piling								
Haul & unload at driving site	load	–	–	–	.10	–	.30	.40
Drive straight	lin ft	.01	–	.04	.05	–	–	.10
Drive battered or angled	lin ft	.02	–	.06	.08	–	–	.16
Hand cut-off	each	–	–	–	.75	–	–	.75
Power cut-off	each	–	–	–	.50	–	–	.50
Steel Shells								
Haul & unload at driving site	load	–	–	–	.03	–	.20	.23
Make-up & drive	lin ft	.02	.02	.03	.06	–	–	.13
Fabricate reinforcing	cwt	–	–	–	–	.60	.05	.65
Place reinforcing	cwt	–	–	–	–	.85	–	.85
Place concrete	cu yd	–	–	–	.50	–	–	.50

Hauling and unloading at driving site units include: Unloading and loading at stockpile on site and hauling to driving site.

Driving units include: Hooking on to pile, pulling into lead and driving. In the case of steel shells, manhours are included for the make-up of sections to obtain desired length.

Cut-off units include: Hand cutting with cross-cut saw or power cutting with chain saw and are based on average cut of 14-inch diameter pile.

Fabricate and place reinforcing units include: Fabricating and placing of reinforcing rods with spirals. This type of reinforcing is usually fabricated in an off-the-job shop and delivered ready to install. If this is the case, eliminate above fabricating charges. If piling is not to be reinforced, eliminate all above reinforcing charges.

Place concrete units include: Placing of ready-mix concrete direct from truck with chutes into piling.

Manhours do not include pile driver set-up or hauling of piles from manufacturer to site. See respective tables for these charges.

STEEL "H" OR "I" BEAM PILING & CAISSONS

MANHOURS PER UNITS LISTED

Item	Units	Manhours						
		Oper. Engr.	Fireman	Pile Driver	Laborer	Iron Worker	Truck Driver	Total
Steel "H" or "I" beams								
Haul & unload at site	ton	.80	—	—	—	1.20	.20	2.20
Driving	lin ft	.02	—	.01	.05	.04	—	.12
Cut-off	each	—	—	—	—	1.10	.01	1.11
Caissons								
Hand excavate	cu yd	—	—	—	4.24	—	—	4.24
Excavate with digger	cu yd	.75	—	—	1.69	—	—	2.44
Wood lagging	mfbm*	—	13.00	—	14.00	—	.30	27.30
Iron rings	cwt	—	1.75	—	1.75	—	.10	3.60
Place concrete	cu yd	—	—	—	3.00	—	—	3.00

Haul and unload units include: Unloading at stockpile and loading, hauling and placing at driving site.

Driving units include: Hooking on pulling into place and driving.

Cut-off units include: Cutting off to desired elevation with torch and hauling away of miscellaneous cut-off pieces.

Excavation units include: Hand or mechanical excavating — manhours do not include trucking of excavated materials — see respective table for this charge.

Woodlagging and ring units include: Fabrication of lagging and placement of lagging and prefabricated rings.

Place concrete units include: Placing of ready-mix concrete direct from trucks with chutes into holes.

Manhours do not include pile driver set-up. See respective table for this charge.

*1,000 foot board measure

PRECAST CONCRETE PILING

MANHOURS REQUIRED FOR UNITS LISTED

Item Description	Unit	Manhours					
		Pile Butt	Laborer	Operating Engineer	Oiler	Truck Driver	Total
Drive							
10" square	LF	0.075	–	0.030	0.015	–	0.120
12" square	LF	0.080	–	0.034	0.017	–	0.131
14" square	LF	0.083	–	0.036	0.018	–	0.137
16" octagonal	LF	0.087	–	0.038	0.019	–	0.144
18" octagonal	LF	0.090	–	0.046	0.023	–	0.159
Cut-Off							
10" square	EA	–	0.500	–	–	0.050	0.055
12" square	EA	–	0.584	–	–	0.050	0.634
14" square	EA	–	0.667	–	–	0.050	0.717
16" octagonal	EA	–	0.750	–	–	0.050	0.800
18" octagonal	EA	–	0.834	–	–	0.050	0.884
Haul and Unload	Ton	–	1.200	0.800	0.400	0.200	2.600

Drive units include hooking on to pile, pulling into leads, and driving.

Cut-Off units include cutting off above dowels to desired elevation with concrete saw and hauling away miscellaneous pieces. Where cut-off is made below the dowels, extra time will be required to reinstall the dowels.

Haul and unload units include unloading at stock pile and loading, hauling and placing at driving site.

Manhours are based on a quantity of 300 piles or more. If less than 300 piles are required increase above manhours 25 percent.

Manhours do not include pile driver set-up. See respective table for this time requirement.

Section 6

FORMWORK

This section covers labor for the complete fabrication, placing and stripping of form work for concrete structures.

Fabrication units include unloading of lumber at fabricating shop or yard, necessary labor for power saw men including pro-rata share of saw filer, fabricating of forms, transporting of fabricated forms to storage location at yard and initial oiling of forms.

Erection units include hauling of fabricated forms and miscellaneous items from yard to erection site within one thousand (1000) feet of storage yard, any field layout done solely by carpenters such as batter board installation, the erection of forms and bracing, the required carpentry clean-up prior to pouring concrete and required carpentry stand-by while concrete is being poured.

Strip and clean units include the stripping and cleaning of forms after initial concrete set, necessary site clean-up and the transporting of forms back to storage yard or new location for re-use.

Units are based on square feet of contact area and re-use and re-oiling factors should be used when applicable instead of initial fabricating units. Complete or major rebuilding of forms should be allowed as an initial fabrication charge.

Units do not include the installation of anchor bolts or miscellaneous embedded items or the installation or dismantling of scaffolding. See respective section and tables for these charges.

FOOTING & HEAVY MAT FORMS

MANHOURS PER SQUARE FOOT

Item	Manhours			
	Carpenter	Laborer	Truck Driver	Total
Continuous Wall Type				
Fabricate	.030	.008	–	.038
Erect	.026	.008	.001	.035
Strip & clean	.008	.015	.001	.024
Total	.064	.031	.002	.097
Spread Type				
Fabricate	.038	.008	–	.046
Erect	.029	.015	.001	.045
Strip & clean	.008	.015	.001	.024
Total	.075	.038	.002	.115
Heavy Mat or Pile Cap				
Fabricate	.035	.010	–	.045
Erect	.054	.022	.001	.077
Strip & clean	.008	.023	.001	.032
Total	.097	.055	.002	.154

Footing form manhours are based on using 2" x 12" material properly braced and anchored. If footing height is to be greater than one foot, add twenty-five percent to above manhours for each additional foot or any fractional part thereof.

Assume heavy mat forms to be made of two-inch material at least three feet high. If mats are to be greater than three feet in height, add ten percent for each additional foot or any part thereof, to above manhours.

Manhours are for initial set-up of one use. If forms are to be re-used, eliminate above manhours for fabrication and apply Re-use Factor manhours on page 103.

Manhours do not include the placement or setting of anchor bolts or miscellaneous embedded steel items. See respective tables for these charges.

Manhours include necessary labor for the various operations as listed above and as outlined in the introduction to this section.

FOUNDATION, TILT-UP WALLS & GRADE BEAMS

MANHOURS PER SQUARE FOOT

Item	Manhours			
	Carpenter	Laborer	Truck Driver	Total
Grade Beams				
Fabricate	.023	.008	—	.031
Erect	.036	.011	.001	.048
Strip & clean	.008	.015	.001	.024
Total	.067	.034	.002	.103
Foundation or Basement Walls				
Fabricate	.033	.007	—	.040
Erect	.042	.011	.001	.054
Strip & Clean	.007	.015	.001	.023
Total	.082	.033	.002	.117
Tilt-Up Walls				
Fabricate	.030	.008	—	.038
Erect	.026	.008	.001	.035
Strip & clean	.008	.015	.001	.024
Total	.064	.031	.002	.097

Manhours are based on the use of plyform material with studs and whalers of two-inch material properly placed and the necessary holes drilled for form ties where required.

Grade beam form manhours are for side forms only, with the assumption that ground will form the bottom of beam.

Tilt-up wall forms are based on being fabricated and placed on level accessible location.

If forms are to be used more than once, eliminate fabrication charges for all but initial use and apply Re-use Factor manhours on page 103.

Manhours do not include the placement or setting of anchor bolts, or miscellaneous embedded steel items. See respective tables for these charges.

Manhours include labor for the various operations as listed above and as outlined in the introduction to this section.

SQUARE WOOD PIER FORMS

MANHOURS PER SQUARE FOOT

Item	Manhours			
	Carpenter	Laborer	Truck Driver	Total
Using 2" x 4" Lumber Clamps				
Fabricate	.029	.018	—	.047
Erect	.035	.018	.001	.054
Strip & clean	.007	.017	.001	.025
Total	.071	.053	.002	.126
Using Metal Clamps				
Fabricate	.029	.018	—	.047
Erect	.035	.018	.001	.054
Strip & clean	006	.017	.001	.024
Total	.070	.053	.002	.125

Manhours are for the use of plywood fabricated forms with the use of 2" x 4" wood clamps or metal clamps as the case may be and includes all necessary bracing and anchoring.

If forms are to be used more than once, eliminate above fabrication manhours for all but initial use and apply Re-use Factor manhours on page 103.

Manhours do not include the placement or setting of anchor bolts or miscellaneous embedded steel. See respective tables for these charges.

FIBER TUBE PIER FORMS

MANHOURS PER LINEAR FOOT

Item	Manhours			
	Carpenter	Laborer	Truck Driver	Total
12" Round				
Fabricate & erect	.045	.008	.001	.054
Strip	.008	.008	.001	.017
Total	.053	.016	.002	.071
14" Round				
Fabricate & erect	.046	.008	.001	.055
Strip	.010	.008	.001	.019
Total	.056	.016	.002	.074
15" Round				
Fabricate & erect	.047	.008	.001	.056
Strip	.012	.008	.001	.021
Total	.059	.016	.002	.077
16" Round				
Fabricate & erect	.048	.008	.001	.057
Strip	.015	.008	.001	.024
Total	.063	.016	.002	.081
18" Round				
Fabricate & erect	.049	.008	.001	.058
Strip	.017	.008	.001	.026
Total	.066	.016	.002	.084

Manhour units cover the complete installation of fiber tubing as listed above, including the fabrication and installation of necessary one- and two-inch planking for bracing and anchoring and the stripping of all items after initial concrete set.

Manhours do not include the placement or setting of anchor bolts or miscellaneous embedded steel items. See respective tables for these charges.

FIBER TUBE PIER FORMS

MANHOURS PER LINEAR FOOT

Item	Manhours			
	Carpenter	Laborer	Truck Driver	Total
20" Round				
Fabricate & erect	.050	.008	.001	.059
Strip	.018	.008	.001	.027
Total	.068	.016	.002	.086
22" Round				
Fabricate & erect	.051	.008	.001	.060
Strip	.020	.008	.001	.029
Total	.071	.016	.002	.089
24" Round				
Fabricate & erect	.052	.008	.001	.061
Strip	.020	.008	.001	.029
Total	.072	.016	.002	.090
26" Round				
Fabricate & erect	.052	.008	.001	.061
Strip	.022	.008	.001	.031
Total	.074	.016	.002	.092
28" Round				
Fabricate & erect	.053	.008	.001	.062
Strip	.022	.008	.001	.031
Total	.075	.016	.002	.093

Manhour units cover the complete installation of fiber tubing as listed above, including the fabrication and installation of necessary one- and two-inch planking for bracing and anchoring and the stripping of all items after initial concrete set.

Manhours do not include the placement or setting of anchor bolts or miscellaneous embedded steel items. See respective tables for these charges.

FIBER TUBE PIER FORMS

MANHOURS PER LINEAR FOOT

Item	Manhours			
	Carpenter	Laborer	Truck Driver	Total
30″ Round				
Fabricate & erect	.054	.008	.001	.063
Strip	.022	.008	.001	.031
Total	.076	.016	.002	.094
32″ Round				
Fabricate & erect	.054	.008	.001	.063
Strip	.023	.008	.001	.032
Total	.077	.016	.002	.095
34″ Round				
Fabricate & erect	.055	.008	.001	.064
Strip	.023	.008	.001	.032
Total	.078	.016	.002	.096
36″ Round				
Fabricate & erect	.056	.008	.001	.065
Strip	.024	.008	.001	.033
Total	.080	.016	.002	.098
38″ Round				
Fabricate & erect	.058	.008	.001	.067
Strip	.024	.008	.001	.033
Total	.082	.016	.002	.100

Manhour units cover the complete installation of fiber tubing as listed above, including the fabrication and installation of necessary one- and two-inch planking for bracing and anchoring and the stripping of all items after initial concrete set.

Manhours do not include the placement or setting of anchor bolts or miscellaneous embedded steel items. See respective tables for these charges.

FIBER TUBE PIER FORMS

MANHOURS PER LINEAR FOOT

Item	Manhours			
	Carpenter	Laborer	Truck Driver	Total
40″ Round				
Fabricate & erect	.060	.008	.001	.069
Strip	.025	.008	.001	.034
Total	.085	.016	.002	.103
42″ Round				
Fabricate & erect	.062	.008	.001	.071
Strip	.025	.008	.001	.034
Total	.087	.016	.002	.105
44″ Round				
Fabricate & erect	.065	.008	.001	.074
Strip	.026	.008	.001	.035
Total	.091	.016	.002	.109
48″ Round				
Fabricate & erect	.070	.008	.001	.079
Strip	.026	.008	.001	.035
Total	.096	.016	.002	.114

Manhour units cover the complete installation of fiber tubing as listed above, including the fabrication and installation of necessary one- and two-inch planking for bracing and anchoring and the stripping of all items after initial concrete set.

Manhours do not include the placement or setting of anchor bolts or miscellaneous embedded steel items. See respective tables for these charges.

METAL WALL FORMS FOR CONTINUOUS FOUNDATIONS

MANHOURS PER SQUARE FOOT

Item	Manhours			
	Carpenter	Laborer	Truck Driver	Total
Foundation Walls to 8' High				
Erect & brace	.053	.021	.002	.076
Strip & clean	.011	.023	.002	.036
Total	.064	.044	.004	.112
Foundation Walls 8' to 12' High				
Erect & brace	.051	.020	.002	.073
Strip & clean	.010	.022	.002	.034
Total	.061	.042	.004	.107

Manhours are based on the installation of prefabricated metal panel forms with the necessary whalers, bracing and anchors made of two-inch materials.

Manhours do not include the installation or setting of anchor bolts or miscellaneous embedded steel items. See respective tables for these charges.

Re-use Factor for re-oiling of forms should be applied for each re-use of above forms. See respective table for this charge.

METAL WALL FORMS FOR CONTINUOUS ABOVE GRADE WALLS

MANHOURS PER SQUARE FOOT

Item	Manhours			
	Carpenter	Laborer	Truck Driver	Total
Walls Ground to 8' High				
Erect & brace	.059	.023	.003	.085
Strip & clean	.012	.031	.003	.046
Total	.071	.054	.006	.131
Walls 8' to 16' High				
Erect & brace	.066	.026	.003	.095
Strip & clean	.014	.035	.003	.052
Total	.080	.061	.006	.147
Walls 16' to 20' High				
Erect & brace	.078	.031	.003	.112
Strip & clean	.017	.041	.003	.061
Total	.095	.072	.006	.173

Manhours are based on the installation of pre-fabricated metal panel forms with the necessary whalers, bracing and anchors made of two-inch materials.

Re-use Factor for re-oiling of forms should be applied for each re-use of above forms. See respective table for this charge.

Manhours do not include scaffolding allowance. If this is necessary, see respective table for this charge.

CONTINUOUS WOOD WALL FORMS

MANHOURS PER SQUARE FOOT

Item	Manhours			
	Carpenter	Laborer	Truck Driver	Total
Walls Ground to 8' High				
Fabricate	.030	.008	–	.038
Erect	.045	.012	.002	.059
Strip & clean	.007	.029	.002	.038
Total	.082	.049	.004	.135
Walls 8' to 16' High				
Fabricate	.030	.008	–	.038
Erect	.051	.013	.002	.066
Strip & clean	.009	.033	.002	.044
Total	.090	.054	.004	.148
Walls 16' to 20' High				
Fabricate	.030	.008	–	.038
Erect	.060	.016	.002	.078
Strip & clean	.010	.039	.002	.051
Total	.100	.063	.004	.167

Manhours are based on the use of plywood panels with two-inch studs and whalers proper-ly placed and the necessary drilled holes for form ties as required.

If forms are to be re-used, eliminate above fabrication manhours and apply Re-use Factor manhours on page 103.

Manhours do not include allowance for scaffolding. If this is necessary, see respective table for this charge.

For battered walls increase manhours twelve (12) percent.

GANG WALL FORMS

MANHOURS PER SQUARE FOOT

	Manhours			
Item	Carpenter	Laborer	Truck Driver	Total
Walls Ground to 16' High				
Fabricate	.027	.007	–	.034
Erect	.054	.014	.002	.070
Strip & clean	.009	.035	.002	.046
Total	.090	.056	.004	.150
Walls 16' to 24' High				
Fabricate	.027	.007	–	0.34
Erect	.062	.016	.002	.080
Strip & clean	.010	.040	.002	.052
Total	.099	.063	.004	.166
Walls 24' to 32' High				
Fabricate	.027	.007	–	.034
Erect	.072	.019	.002	.093
Strip & clean	.012	.046	.002	.060
Total	.111	.072	.004	.187

Manhours are based on the use of eight-foot high plywood panels with two-inch studs and whalers properly placed. including double studs at lifting points. and necessary drilled holes for form ties as required.

Manhours are based on one use of forms. Where forms are to be re-used. eliminate above fabrication manhours and add re-use factor manhours on page 103.

Manhours include fabrication. hauling form materials to and from the job fabrication and storage yards. wall layout. setting form. connecting. aligning. bracing. stripping. and moving.

Manhours exclude rig operator for lifting forms. Add for this operation as required.

Where patterned architectural finish is required increase manhours twenty (20) percent for placement cf additional forming material and handling.

For battered gang wall forming increase manhours twelve (12) percent.

SQUARE WOOD COLUMN FORMS

MANHOURS PER SQUARE FOOT

Item	Manhours			
	Carpenter	Laborer	Truck Driver	Total
Exterior Columns				
Fabricate	.029	.013	—	.042
Erect	.052	.026	.001	.079
Strip & clean	.010	.026	.001	.037
Total	.091	.065	.002	.158
Interior Columns				
Fabricate	.036	.013	—	.049
Erect	.065	.033	.001	.099
Strip & clean	.010	.026	.001	.037
Total	.111	.072	.002	.185

Manhours include all necessary labor for the actual fabrication, erecting and stripping of exterior and interior columns as outlined above using plywood side with one and two inch planking for necessary clamps, bracing and anchoring.

If forms are to be re-used. above fabrication manhours should be eliminated for all but initial use and Re-use Factor manhours on page 103 should be substituted in their place.

Manhours do not include scaffolding. If this is necessary, see respective table for this charge.

FIBER TUBE COLUMN FORMS

MANHOURS PER LINEAR FOOT

Item	Manhours			
	Carpenter	Laborer	Truck Driver	Total
12" Round				
Fabricate & erect	.036	.006	.001	.043
Strip	.006	.006	.001	.013
Total	.042	.012	.002	.056
14" Round				
Fabricate & erect	.036	.006	.001	.043
Strip	.008	.006	.001	.015
Total	.044	.012	.002	.058
15" Round				
Fabricate & erect	.037	.006	.001	.044
Strip	.009	.006	.001	.016
Total	.046	.012	.002	.060
16" Round				
Fabricate & erect	.038	.006	.001	.045
Strip	.012	.006	.001	.019
Total	.050	.012	.002	.064
18" Round				
Fabricate & erect	.039	.006	.001	.046
Strip	.013	.006	.001	.020
Total	.052	.012	.002	.066

Manhours include time limitations for the complete fabrication and erection of one- and two-inch materials for bracing and anchoring as well as the erection of all fiber tubing, bracing and the removal of anchoring and stripping of all components after initial concrete set.

Manhours do not include scaffolding. If this is necessary, see respective table for this charge.

FIBER TUBE COLUMN FORMS

MANHOURS PER LINEAR FOOT

Item	Manhours			
	Carpenter	Laborer	Truck Driver	Total
20" Round				
Fabricate & erect	.040	.006	.001	.047
Strip	.014	.006	.001	.021
Total	.054	.012	.002	.068
22" Round				
Fabricate & erect	.041	.006	.001	.048
Strip	.016	.006	.001	.023
Total	.057	.012	.002	.071
24" Round				
Fabricate & erect	.042	.006	.001	.049
Strip	.016	.006	.001	.023
Total	.058	.012	.002	.072
26" Round				
Fabricate & erect	.043	.006	.001	.050
Strip	.018	.006	.001	.025
Total	.061	.012	.002	.075
28" Round				
Fabricate & erect	.044	.006	.001	.051
Strip	.018	.006	.001	.025
Total	.062	.012	.002	.076

Manhours include time limitations for the complete fabrication and erection of one- and two-inch materials for bracing and anchoring as well as the erection of all fiber tubing, bracing and the removal of anchoring and stripping of all components after initial concrete set.

Manhours do not include scaffolding. If this is necessary, see respective table for this charge.

FIBER TUBE COLUMN FORMS

MANHOURS PER LINEAR FOOT

Item	Manhours			
	Carpenter	Laborer	Truck Driver	Total
30" Round				
Fabricate & erect	.045	.006	.001	.052
Strip	.018	.006	.001	.025
Total	.063	.012	.002	.077
32" Round				
Fabricate & erect	.046	.006	.001	.053
Strip	.019	.006	.001	.026
Total	.065	.012	.002	.079
34" Round				
Fabricate & erect	.047	.006	.001	.054
Strip	.020	.006	.001	.027
Total	.067	.012	.002	.081
36" Round				
Fabricate & erect	.048	.006	.001	.055
Strip	.021	.006	.001	.028
Total	.069	.012	.002	.083
38" Round				
Fabricate & erect	.050	.006	.001	.057
Strip	.021	.006	.001	.028
Total	.071	.012	.002	.085

Manhours include time limitations for the complete fabrication and erection of one- and two-inch materials for bracing and anchoring as well as the erection of all fiber tubing, bracing and the removal of anchoring and stripping of all components after initial concrete set.

Manhours do not include scaffolding. If this is necessary, see respective table for this charge.

FIBER TUBE COLUMN FORMS

MANHOURS PER LINEAR FOOT

Item	Manhours			
	Carpenter	Laborer	Truck Driver	Total
40" Round				
Fabricate & erect	.053	.006	.001	.060
Strip	.022	.006	.001	.029
Total	.075	.012	.002	.089
42" Round				
Fabricate & erect	.055	.006	.001	.062
Strip	.022	.006	.001	.029
Total	.077	.012	.002	.091
44" Round				
Fabricate & erect	.059	.006	.001	.066
Strip	.023	.006	.001	.030
Total	.082	.012	.002	.096
48" Round				
Fabricate & erect	.064	.006	.001	.071
Strip	.024	.006	.001	.031
Total	.088	.012	.002	.102

Manhours include time limitations for the complete fabrication and erection of one- and two-inch materials for bracing and anchoring as well as the erection of all fiber tubing, bracing and the removal of anchoring and stripping of all components after initial concrete set.

Manhours do not include scaffolding. If this is necessary, see respective table for this charge.

STEEL COLUMN CAPITALS
FOR ROUND COLUMNS

MANHOURS REQUIRED EACH

Column Capital Diameter At Top	Manhours				
	Carpenter	Laborer	Operator Engineer	Truck Driver	Total
3'6"	1.250	1.250	0.835	0.584	3.919
4'0"	1.434	1.434	0.835	0.584	4.287
4'6"	1.617	1.617	0.835	0.584	4.653
5'0"	1.800	1.800	0.835	0.584	5.019
5'6"	1.984	1.984	0.835	0.584	5.387
6'0"	2.167	2.167	0.835	0.584	5.753
6'6"	2.350	2.350	0.835	0.584	6.119
7'0"	2.517	2.517	0.835	0.584	6.453

Manhours include handling. job hauling. installing. bracing. stripping. and cleaning of steel column capitals on round fiber tube forms up to a height of 16 feet above floor elevation.

For heights greater than 16 feet increase carpenter. laborer. and operator engineer time .034 manhours each for each additional linear foot of height.

Manhours do not include placement of embedded items or scaffolding. If this is necessary. see respective tables for these time frames.

RADIAL & RETAINING WOOD WALL FORMS

MANHOURS PER SQUARE FOOT

Item	Manhours			
	Carpenter	Laborer	Truck Driver	Total
Radial Walls to 4' High				
Fabricate	.049	.010	—	.059
Erect	.064	.017	.001	.082
Strip & clean	.010	.022	.001	.033
Total	.123	.049	.002	.174
Radial Walls 4' to 12' High				
Fabricate	.059	.012	—	.071
Erect	.077	.020	.002	.099
Strip & clean	.012	.026	.002	.040
Total	.148	.058	.004	.210
Retaining Wall to 10' High				
Fabricate	.036	.007	—	.043
Erect	.047	.013	.001	.061
Strip & clean	.007	.016	.001	.024
Total	.090	.036	.002	.128

Manhours are based on the use of plywood panels with two-inch studs and whalers properly placed and the necessary drilled holes for form ties as required.

If forms are to be used more than once. eliminate above fabrication manhours and apply Re-use Factor manhours on page 103 for all except initial use.

WOOD BEAM & GIRDER FORMS

MANHOURS PER SQUARE FOOT

Item	Manhours			
	Carpenter	Laborer	Truck Driver	Total
Square Beams				
Fabricate	.038	.011	—	.049
Erect	.071	.026	.001	.098
Strip & clean	.008	.023	.001	.032
Total	.117	.060	.002	.179
Shored Beam & Girder				
Fabricate	.032	.013	—	.045
Erect	.088	.035	.001	.124
Strip & clean	.017	.045	.001	.063
Total	.137	.093	.002	.232
Hung from Steelbeam & Girder				
Fabricate	.027	.011	—	.038
Erect	.073	.029	.001	.103
Strip & clean	.014	.037	.001	.052
Total	.114	.077	.002	.193

Manhours are based on the use of plywood and one- and two-inch planking materials all properly braced and anchored.

Sufficient time has been allowed to cover the above outlined operations and necessary items of labor operations as covered by the introduction to this section.

If forms are to be re-used, apply Re-use Factors on page 103 in place of above listed fabrication manhours for all but initial use.

Manhours do not include scaffolding. If this is necessary, see respective table for this charge.

COLUMN DROP HEADS & BOX OUT FOR OPENING FORMS

MANHOURS PER SQUARE FOOT

Item	Manhours			
	Carpenter	Laborer	Truck Driver	Total
Column Drop Heads				
Fabricate & erect	.11	.06	.01	.18
Strip & clean	.01	.03	.01	.05
Total	.12	.09	.02	.23
Box Out for Openings				
Substructure				
Fabricate & place	.08	.03	—	.11
Strip	.01	.02	.01	.04
Total	.09	.05	.01	.15
Box Out for Openings				
Superstructure				
Fabricate & place	.09	.04	—	.13
Strip	.01	.03	.02	.06
Total	.10	.07	.02	.19

Column drop head manhours are based on depression in slab at column head of four to eight inches deep and five feet by seven feet square.

Sub-structure box out for opening manhours are average for below ground work and super-structure manhours are average for above ground work.

Manhours include time only for operations as outlined above and as described in the introduction to this section.

Manhours are based on the use of one- or two-inch material for box out openings and plyform sheathing for dropheads.

SHEET METAL BOX-OUT FORMS AND SLEEVES

MANHOURS REQUIRED EACH

Item	Manhours			
	Carpenter	Laborer	Truck Driver	Total
Box-Out Forms on Deck				
Sizes 2"x6" through 12"x20"	.334	.250	.100	.684
Sleeves				
Sizes 1½" through 24" Round				
Set on deck forms	.334	.084	.050	.468
Set of hand set wall forms	.500	.100	.050	.650
Set on gang wall forms	.667	.167	.050	.884
Set on beam forms	.500	.084	.050	.634

Manhours include checking out of job storage. handling. hauling, and installing in position as outlined above.

For installation of regular pipe sleeves. see "Miscellaneous Embedded Items."

Manhours do not include placement of forms or other concrete items. See respective tables for these time frames. frames.

WOOD FORMS FOR GROUND FLOOR SLABS

MANHOURS PER SQUARE FOOT

Item	Manhours			
	Carpenter	Laborer	Truck Driver	Total
Slab to 6" Thick				
Fabricate	.027	.007	—	.034
Erect	.024	.007	.001	.032
Strip & clean	.008	.014	.001	.023
Total	.059	.028	.002	.089
Slab to 12" Thick				
Fabricate	.030	.008	—	.038
Erect	.026	.008	.001	.035
Strip & clean	.008	.015	.001	.024
Total	.064	.031	.002	.097

Slab on ground form manhours are based on the complete fabrication and erection of two-inch materials including all necessary bracing and anchoring.

Manhours include necessary labor for the various operations as listed above and as outlined in the introduction to this section.

Manhours are for initial set-up of one use. If forms are to be re-used, eliminate above manhours for fabrication and apply Re-use Factor manhours on page 103.

Manhours do not include the placement or setting of screeds equipment pad forms, anchor bolts, or miscellaneous embedded steel items. See respective tables for these charges.

WOOD FORMS FOR ELEVATED SLABS

MANHOURS PER SQUARE FOOT

Item	Manhours			
	Carpenter	Laborer	Truck Driver	Total
Flat Slab with Wood Shores				
Make-up & erect	.067	.036	.001	.104
Strip & clean	.004	.036	.001	.041
Total	.071	.072	.002	.145
Flat Slab with Adjustable Shores				
Make-up & erect	.043	.030	.001	.074
Strip & clean	.004	.036	.001	.041
Total	.047	.066	.002	.115
Flat Slab Hung from Steel				
Fabricate & erect	.041	.009	.001	.051
Strip & clean	.005	.020	.001	.026
Total	.046	.029	.002	.077

Manhours are based on the use of plywood and properly sized timbers as required for bracing and supporting.

Sufficient time has been allowed to cover the above outlined operations and necessary items of labor operations as covered by the introduction to this section.

If forms are to be re-used. apply Re-use Factor manhours on page 103 in place of above listed fabrication manhours for all but initial use.

Manhours do not include the placement or setting of screeds, equipment pad forms, anchor bolts, miscellaneous embedded steel items or scaffolding. See respective tables for these items.

METAL FLOOR PAN FORMS

MANHOURS PER SQUARE FOOT

Item	Manhours			
	Carpenter	Laborer	Truck Driver	Total
Metal Pan with Adjustable Shores				
Cut & erect lumber	.040	.023	.002	.065
Erect pans & shores	.010	.015	.002	.027
Strip & clean	—	.018	.002	.020
Total	.050	.056	.006	.112
Metal Pan with Wood Shores				
Cut & erect lumber	.050	.040	.002	.092
Erect pans & shores	.010	.015	.002	.027
Strip & clean	—	.018	.002	.020
Total	.060	.073	.006	.139

Manhours are based on the installation of prefabricated metal pan forms with the necessary size timber supports.

Sufficient time has been allowed to cover the above outlined operations and necessary items of labor operations as covered by the introduction to this section.

Manhours do not include the placement or setting of screeds equipment pad forms, anchor bolts, miscellaneous embedded steel items or scaffolding. If required, see respective tables for these items.

WOOD TRENCH & CURB FORMS

MANHOURS PER UNITS LISTED

Item	Unit	Manhours			
		Carpenter	Laborer	Truck Driver	Total
Trench Forms					
Fabricate	sq ft	.03	.02	—	.05
Erect	sq ft	.06	.04	.01	.11
Strip & clean	sq ft	.01	.04	.01	.06
Total	sq ft	.10	.10	.02	.22
Curb Forms on Ground					
Build in place	lin ft	.07	.03	.01	.11
Strip & clean	lin ft	.01	.02	.01	.04
Total	lin ft	.08	.05	.02	.15
Curb Forms Elevated					
Make-up & erect	lin ft	.11	.02	.01	.14
Strip & clean	lin ft	.01	.03	.01	.05
Total	lin ft	.12	.05	.02	.19

Trench Form manhours are based on slab construction of two-inch material with wall forms of plyform material, all properly braced and anchored.

Curb Form manhours are based on the use of two-inch material up to one foot in height, properly braced and anchored.

These manhours cover necessary time for the various operations as listed above and as outlined in the introduction to this section.

STAIRWELL & STAIR FORMS

MANHOURS PER UNITS LISTED

Item	Unit	Manhours			
		Carpenter	Laborer	Truck Driver	Total
Stairwell					
Fabricate	sq ft	.04	.02	—	.06
Erect	sq ft	.08	.04	.01	.13
Strip & clean	sq ft	.02	.04	.01	.07
Total	sq ft	.14	.10	.02	.26
Straight Stair Forms					
Fabricate, erect & strip	lin ft of rise	.20	.07	.01	.28
Small Step Forms					
Fabricate & erect	sq ft	.14	.07	.01	.22
Strip	sq ft	.01	.05	—	.06
Total	sq ft	.15	.12	.01	.28
Hand or Guard Rail Including Concrete Posts					
Fabricate & erect	lin ft	.17	.10	.01	.28
Strip	lin ft	.02	.06	—	.08
Total	lin ft	.19	.16	.01	.36

Above manhours for the units listed are based on the use of one- and two-inch planking and plywood, all as may apply to the above type forming and all properly braced, anchored and supported.

All the above described operations, plus those listed in the introduction to this section, have been given due consideration.

No allowance has been considered for the use of scaffolding. If necessary, see respective table for this charge.

LOADING DOCK, PLATFORMS & CANOPY WOOD FORMS

MANHOURS PER SQUARE FOOT

Item	Manhours			
	Carpenter	Laborer	Truck Driver	Total
Loading Dock & Ramp				
Fabricate	.05	.02	—	.07
Erect	.07	.02	.01	.10
Strip & clean	.01	.03	.01	.05
Total	.13	.07	.02	.22
Platforms				
Make-up & erect	.12	.06	—	.18
Strip & clean	.02	.04	.02	.08
Total	.14	.10	.02	.26
Canopy				
Fabricate	.05	.01	—	.06
Erect	.13	.03	.01	.17
Strip & clean	.01	.05	.01	.07
Total	.19	.09	.02	.30

Manhours are based on the use of one- and two-inch sheathing and plywood as may be necessary for the above type forming, all properly braced, anchored and supported.

Manhours do not include scaffolding as may be necessary. See respective table for this charge.

If forms are to be used more than once. eliminate fabrication charges for all but initial use and apply Re-use Factor manhours on page 103.

SILLS, LINTELS, COPING & BALCONY
FORM WOOD

MANHOURS PER SQUARE FOOT

Item	Manhours			
	Carpenter	Laborer	Truck Driver	Total
Sills Poured in Place				
Make-up & erect	.11	.09	—	.20
Strip & clean	.02	.04	.01	.07
Total	.13	.13	.01	.27
Lintels Poured in Place				
Make-up & erect	.14	.11	—	.25
Strip & clean	.02	.05	.01	.08
Total	.16	.16	.01	.33
Coping Poured in Place				
Make-up & Erect	.12	.10	—	.22
Strip & clean	.02	.04	.01	.07
Total	.14	.14	.01	.29
Balconies & Ornamental Projections				
Make-up & erect	.23	.02	—	.25
Strip & clean	.06	.08	.01	.15
Total	.29	.10	.01	.40

Manhours for sills, lintels and coping are based on the fabrication and erection in place of one- and two-inch material, all properly braced, anchored and supported.

Manhours for balconies and ornamental projections of one- and two-inch sheathing and plywood, along with minor prefabricated furnished sheet metal shapes, have been given due consideration in the above table.

For additional information as to manhour coverage, see introduction to this section.

COLUMN & BEAM POCKETS AND ENCASEMENTS, CHAMFER, SCREEDS & KEYWAYS

MANHOURS PER UNITS LISTED

Item	Unit	Manhours			
		Carpenter	Laborer	Truck Driver	Total
Column & Beam Pocket Forms					
Fabricate & erect	each	.45	.15	–	.60
Strip	each	.01	.20	.01	.22
Total	each	.46	.35	.01	.82
Steel Column & Beam Encasement					
Fabricate	sq ft	.05	.02	–	.07
Erect	sq ft	.06	.03	.01	.10
Strip & clean	sq ft	.02	.04	.01	.07
Total	sq ft	.13	.09	.02	.24
Chamfer Strip	100 lin ft	1.5	–	–	1.5
Set Screeds	100 lin ft	1.6	–	–	1.6
Keyway Strip	100 lin ft	1.5	–	–	1.5

Manhours for column and beam pocket and fireproof encasement forms are based on the use of one- and two-inch planks and plywood as may be necessary for these operations.

Manhours for Chamfer and Keyway strips include necessary time allowance for angle cutting.

See introduction to this section for additional operations covered in above manhours.

EQUIPMENT FOUNDATIONS – SIMPLE LAYOUT

MANHOURS PER SQUARE FOOT

Item	Manhours			
	Carpenter	Laborer	Truck Driver	Total
Square Pads 6" to 18" High Ground Floor				
Build in place	.14	.03	.01	.18
Strip & clean	.01	.03	–	.04
Total	.15	.06	.01	.22
Square Pads 6" to 18" High Elevated Floors				
Build in place	.16	.03	.01	.20
Strip & clean	.01	.04	–	.05
Total	.17	.07	.01	.25
Square Pads to 4' High Ground Floor				
Fabricate & erect	.17	.04	.01	.22
Strip & clean	.02	.04	–	.06
Total	.19	.08	.01	.28
Square Pads to 4' High Elevated Floors				
Fabricate & erect	.19	.04	.01	.24
Strip & clean	.02	.05	–	.07
Total	.21	.09	.01	.31

Manhours are based on the fabrication and installation of two-inch materials for formwork to eighteen inches high, and plywood sheathing for forms to four feet high, all properly braced and anchored in place.

A simple layout is that of a small square pad poured either integral with floor or over pre-set dowels left purposely in pre-poured floor for this reason.

Manhours do not include the placement or setting of anchor bolts or miscellaneous embedded steel items. See respective tables for these charges.

EQUIPMENT FOUNDATIONS – COMPLEX LAYOUT

Bulky, Offset, Skewed and Angled

MANHOURS PER SQUARE FOOT

Item	Manhours			
	Carpenter	Laborer	Truck Driver	Total
Average All Heights & Sizes				
Fabricate & erect	.20	.08	.02	.30
Strip & clean	.05	.12	–	.17
Total	.25	.20	.02	.47
Tank Cradle Forms				
Build in place	.13	.04	.01	.19
Strip & clean	.01	.03	.01	.04
Total	.14	.07	.02	.23

Complex Foundation manhours are average for all sizes and shapes and are based on the use of one- and two-inch planking, plywood sheathing and minor sheet metal cuts and bends.

A complex layout is that of a large and bulky foundation with many offsets, skews and angles, such as a foundation for a turbo-generator, etc.

Manhours do not include the placement or setting of anchor bolts or miscellaneous embedded steel items. See respective tables for these charges.

RE-USE FACTORS

Item	Manhours		
	Carpenter	Laborer	Total
Repairs — First Re-Use	.01	.003	.013
Repairs — Second Re-Use	.02	.004	.024
Repairs — Third Re-Use	.03	.006	.036
Repairs — Subsequent Re-Uses	.04	.007	.047
Oiling after each use	—	.005	.005

Re-use Factor Manhours are average for the minor replacement or repairs to all types of wood forming.

The above Re-use Factor manhours should be substituted for the initial fabrication manhours for all form work, in accordance with its estimated re-uses.

If major repairs are required for re-use of form work, it should be charged at initial fabrication for that particular type of form work.

Section 7

REINFORCING STEEL AND MESH

Included in this section are manhour tables for the installation of reinforcing rods and welded wire mesh.

Manhours listed are those of iron workers or rod busters and include the placement of all accessories as may be required.

Crane operator for hoisting materials and truck driver for job hauling materials are not included in the manhours listed. The estimator should give individual consideration to these operations if required.

Manhours listed are for five (5) tons or more of rods and ten thousand (10,000) square feet or more of mesh. If less than these quantities are required, the manhour tables should be increased twenty (20) percent for rods and twenty-five (25) percent for mesh.

REINFORCING RODS

MANHOURS PER UNITS LISTED

Item	Manhours			
	Fabricate Cut and Bend		Place and Tie	
	(per ton)	(per cwt)	(per ton)	(per cwt)
Unload, sort, and pile rods	1.75	0.088	–	–
Continuous footings	5.12	0.256	15.50	0.775
Spread type footings	5.18	0.259	15.70	0.785
Heavy mat or pile cap	4.63	0.232	13.23	0.662
Pile cap tie beams	4.80	0.240	14.55	0.728
Grade beams	5.32	0.266	14.79	0.740
Ground slabs	8.50	0.425	25.00	1.250
Round columns	11.67	0.584	25.36	1.268
Square columns	9.44	0.472	22.47	1.124
Structural walls	12.86	0.643	29.91	1.496
Flat elevated slabs	20.20	1.010	45.90	2.295
Inside elevated beams	10.49	0.525	23.84	1.192
One and two way beams and slabs	21.37	1.069	46.46	2.323
One way joist (pan construction)	26.19	1.310	56.93	2.847
Waffle flat slabs and joist ribs	19.24	0.962	41.82	2.091
Equipment foundations—simple	13.25	0.663	28.81	1.441
Equipment foundations—complex	16.59	0.830	34.57	1.729
Average—all types	11.56	0.578	28.43	1.422

Unload, sort, and pile units include the unloading of flat cars or trucks at the yard and stock piling by size and length.

Fabricate, cut, and bend units include transporting of materials from stock pile to fabrication site, cutting, bending, tagging, any required transporting of fabricated items to storage, and associated clean-up of fabrication site.

Place and tie units include transporting of fabricated materials to within one thousand (1000) feet of erection site, the necessary field lay-out, placement of any spacers, supports, etc., placing and tying of rods as specified, stand-by time while concrete is being placed, and associated clean-up of area.

Average all types units include average manhours for all above operations.

REINFORCING MESH

MANHOURS PER UNITS LISTED

Mesh Size	Manhours	
	Per CSF	Per SF
6"x6"−3/3	0.54	0.0054
6"x6"−4/4	0.45	0.0045
6"x6"−6/6	0.45	0.0045
6"x6"−8/8	0.43	0.0043
6"x6"−10/10	0.40	0.0040
4"x4"−4/4	0.47	0.0047
4"x4"−8/8	0.50	0.0050
4"x4"−10/10	0.54	0.0054

Above manhours include transporting mesh from storage to within one thousand (1000) feet of erection site, the necessary field layout, placement of any required spacers, supports, etc., cutting and placing of mesh as specified, stand-by and positioning time while concrete is being placed, and associated clean-up of area.

Manhours do not include placement of reinforcing rods. See respective table for this time requirement.

Section 8

MISCELLANEOUS EMBEDDED ITEMS

Included in this section are manhour tables for the installation of items to be embedded in concrete.

The manhours that follow are an average of many projects of varied nature. They include handling and hauling of prefabricated items from storage yard or warehouse to within one thousand (1000) feet of erection site and installation of items.

Units do not include time for flat sketch man or field engineer, if these items are required and are to be charged to the direct cost, they must be added.

HOOK TYPE ANCHOR BOLT INSTALLATION

MANHOURS EACH

Size	Manhours Each For Overall Length							
	0' 8"	1' 0"	1' 6"	2' 0"	2' 6"	3' 0"	3' 6"	4' 0"
1/4"	.15	.15	.20	.20	—	—	—	—
3/8"	.15	.15	.20	.22	—	—	—	—
1/2"	.15	.15	.25	.28	—	—	—	—
5/8"	.15	.20	.25	.28	.30	.33	—	—
3/4"	.18	.20	.28	.30	.35	.38	—	—
7/8"	—	—	.40	.43	.45	.48	.50	.53
1"	—	—	.40	.45	.48	.50	.53	.58
1-1/4"	—	—	.48	.50	.50	.53	.55	.65
1-1/2"	—	—	.50	.55	.55	.58	.60	.70
1-3/4"	—	—	.55	.58	.60	.65	.68	.73
2"	—	—	—	.65	.68	.70	.75	.78
2-1/4"	—	—	—	.70	.73	.75	.78	.80
2-1/2"	—	—	—	.75	.78	.78	.80	.85

Manhours are based on overall length of anchor bolt from end to end including hook and are average for all heights.

Manhours are for installation of template and bolt, or bolt and sleever, as the case may be, for the size and length as outlined above.

All bolts 7/8" and larger are assumed to be sleeved and those smaller than 7/8" round are assumed to be without sleeves.

When converting manhours to labor dollars, consideration should be given to the type of crew or crafts involved in the installation of above anchor bolts and a crew composite rate applied accordingly.

Manhours do not include fabrication of bolts.

Manhours do not include field engineering time spent aligning and checking bolts. This is usually a part of field overhead and should be considered as such.

For sizes not listed, take the next highest listing.

HOOK TYPE ANCHOR BOLT INSTALLATION

MANHOURS EACH

Size	Manhours each For Overall Length							
	4' 6"	5' 0"	5' 6"	6' 0"	6" 6"	7' 0"	7' 6"	8' 0"
7/8"	.90	.98	1.10	1.15	1.23	1.30	1.35	1.40
1"	.93	1.00	1.15	1.25	1.28	1.38	1.40	1.54
1-1/4"	.95	1.10	1.25	1.28	1.30	1.40	1.45	1.50
1-1/2"	.98	1.15	1.28	1.33	1.38	1.43	1.48	1.58
1-3/4"	1.00	1.25	1.33	1.40	1.43	1.50	1.55	1.60
2"	1.10	1.28	1.40	1.45	1.50	1.58	1.60	1.65
2-1/4"	1.15	1.33	1.45	1.48	1.55	1.60	1.63	1.68
2-1/2"	1.25	1.40	1.48	1.53	1.60	1.63	1.68	1.70

Manhours are based on overall length of anchor bolt from end to end including hook and are average for all heights.

Manhours are for the installation of template and bolt and sleeve for the size and length as outlined above.

All bolts listed above are assumed to be sleeved.

When converting above manhours to labor dollars, consideration should be given to the type of crew or crafts involved in the installation of above anchor bolts and a crew composite rate applied accordingly.

Manhours do not include fabrication of bolts.

Manhours do not include field engineering time spent aligning and checking anchor bolts. This is usually a part of field overhead and should be considered as such.

For sizes not listed, take the next highest listing.

INSTALLATION OF STRAIGHT TYPE ANCHOR BOLTS

MANHOURS EACH

Size	Manhours Each for Overall Length					
	0' 8"	1' 0"	1' 6"	2' 0"	2' 6"	3' 0"
1/4"	.10	.12	.15	.18	.20	.25
3/8"	.10	.12	.18	.20	.22	.28
1/2"	.10	.15	.18	.20	.25	.30
5/8"	.15	.15	.20	.22	.28	.33
3/4"	.15	.18	.20	.25	.30	.38
1"	.15	.18	.23	.25	.33	.40

LOOPS AND SCREW ANCHORS

MANHOURS EACH

Item	Manhours
Coil Loops—½"x4" through 1½"x12"	0.15
Screw anchors—½" through 1½"	0.12

Manhours are based on the installation of template and bolt for the size and length as outlined above and are average for all heights.

If a mixed crew of various crafts is used in the setting of above bolts, consideration should be given this when arriving at a composite rate for the conversion of manhours to labor dollars.

Manhours do not include engineering time spent in the aligning or checking of bolts. This is usually a part of field overhead and should be considered as such.

For sizes not listed, take the next highest listing.

ANGLES, SLEEVES, INSERTS, SLOTS & UNISTRUT

MANHOURS PER UNITS LISTED

Item	Unit	Manhours		
		Carpenter	Truck Driver	Total
Pipe Sleeves through 12"				
Diameter				
Substructure	each	.600	.020	.620
Superstructure	each	.700	.030	.730
Pipe Sleeves through 36"				
Diameter				
Substructure	each	.750	.030	.780
Superstructure	each	.850	.050	.900
Inserts				
Substructure	each	.120	.001	.121
Superstructure	each	.150	.002	.152
Anchor Slot & Unistrut				
Substructure	lin ft	.010	.001	.011
Superstructure	lin ft	.017	.001	.018
Embedded Angle				
Substructure	pound	.020	.001	.021
Superstructure	pound	.031	.002	.033

Manhours are average for the above described items and include handling, hauling and installing.

If project is to be constructed in an area where craft jurisdiction differs from that as shown above, the correct craft should be substituted and the same manhours as appears above applied.

CONCRETE

It is the intent of this section to cover the complete installation of concrete in pre-erected forms and with pre-erected reinforcing steel if required.

Chuting Units include the erection of minor prefabricated wood chutes and supports. If the operation consists of chuting concrete direct from truck with attached steel chutes, carpentry manhours throughout the tables should be eliminated.

Buggy units include the moving of hand powered, three to six cubic feet buggies a distance up to two hundred (200) feet. If distance is to be greater than two hundred feet add .05 manhours for each additional fifty (50) linear feet. If power buggies are to be used instead of hand type, decrease the manhours throughout the following tables for this operation by twenty (20) percent.

Conveyor and buggy units include the transporting of concrete into a preset hopper and discharging into buggies for placement into forms, buggy portions of these units have been given due consideration as outlined above for buggy units.

Crane, bucket and buggy units include the placing of concrete from truck or mixer into a one-half to two cubic yard bucket, the raising and positioning of bucket by a crane, the discharge of concrete from bucket into a preset hopper which is to be discharged into buggies for placement into forms. Buggy portion of manhour time for this type installation is similar to that as outlined above for buggy units.

Hoist and buggy units include the hoisting of pre-mixed concrete in bulk, dumping into hopper which will discharge into buggies for transporting and dumping into forms. Manhours for buggy portion of this type operation are as outlined under buggy units. It is not the intent of the hoist units to suffice for an elevator type hoist capable of moving loaded buggies from ground to elevated areas. If this is required, additional manhours must be added for this operation.

106

Crane and bucket units include the placement of concrete directly into forms from one-half to two cubic yard bucket swung into position by the use of a crane.

The individual placing units as appear under this section do not include the fabrication of chutes, hoists or platforms or finishing. These items are covered under separate operational tables in this section.

FOOTINGS

MANHOURS PER CUBIC YARD

Item	Manhours				
	Laborer	Carpenter	Oper. Engr.	Oiler	Total
Continuous Wall Type Footings					
Chute	.46	.03	—	—	.49
Buggies	.60	—	—	—	.60
Crane & bucket	.81	—	.03	.03	.87
Conveyor & buggy	.70	—	.03	—	.73
Spread Type Footings					
Chute	.53	.04	—	—	.57
Buggies	.70	—	—	—	.70
Crane & bucket	.88	—	.03	.03	.94
Conveyor & buggy	.79	—	.03	—	.82
Heavy Mat or Pile Cap					
Chute	.35	.05	—	—	.40
Buggies	.75	—	—	—	.75
Crane & bucket	.38	—	.06	.06	.50

Manhours include complete placement and vibration of concrete for the above items all in accordance with the introduction to this section.

Manhours do not include necessary craft stand-by time, fabrication of special ramps or finishing of concrete. See respective tables for these charges.

FOUNDATION WALLS & GRADE BEAMS

MANHOURS PER CUBIC YARD

Item	Manhours				
	Laborer	Carpenter	Oper. Engr.	Oiler	Total
Grade Beams					
Chute	.55	.04	—	—	.59
Buggies	.91	—	—	—	.91
Crane & bucket	.94	—	.03	.03	1.00
Conveyor & buggy	.91	—	.07	—	.98
Foundation or Basement Walls					
Chute	.55	.04	—	—	.59
Buggies	1.00	—	—	—	1.00
Crane & bucket	.99	—	.04	.04	1.07
Conveyor & buggy	1.00	—	.07	—	1.07

Above manhours include necessary set-up, placement and vibration of concrete all in accordance with the introduction to this section.

Manhours do not include craft stand-by time, fabrication of special ramps or finishing of concrete. See respective tables for these charges.

PIERS

MANHOURS PER CUBIC YARD

Item	Manhours				
	Laborer	Carpenter	Oper. Engr.	Oiler	Total
Square Piers					
Chute	.70	.05	—	—	.75
Buggies	.96	—	—	—	.96
Crane & bucket	1.02	—	.06	.06	1.14
Conveyor & buggy	.96	—	.10	—	1.06
Round Piers					
Chute	.74	.05	—	—	.79
Buggies	1.00	—	—	—	1.00
Crane & bucket	1.07	—	.06	.06	1.19
Conveyor & buggy	1.00	—	.11	—	1.11

Manhours include set-up time and the placement and vibration of concrete for square and round piers as described above and in accordance with the introduction to this section.

Additional allowance has been given to the placement of concrete in round piers due to the fact that additional care must be taken in this type of formwork.

Manhours do not include craft stand-by time, fabrication and installation of special ramps or finishing of concrete. See respective tables for these charges.

COLUMNS

MANHOURS PER CUBIC YARD

Item	Manhours			
	Laborer	Oper. Engr.	Oiler	Total
Exterior Square Columns				
Hoist & buggies	1.43	.14	–	1.57
Crane, bucket & buggies	.99	.08	.08	1.15
Interior Square Columns				
Hoist & buggies	1.79	.16	–	1.95
Crane, bucket & buggies	1.24	.10	.10	1.44
Round Columns (Interior)				
Hoist & buggies	1.87	.17	–	2.04
Crane bucket & buggies	1.30	.11	.11	1.52
Column Caps				
Hoist & buggies	2.40	.15	–	2.55
Crane, bucket & buggies	1.95	.12	.12	2.19

Manhours are for the pouring and vibrating of above types of columns in accordance with the introduction of this section.

Additional allowance has been given to the placement of round columns due to the fact that additional care must be taken in this type of form work.

Manhours do not include craft stand-by time, the fabrication, installation and dismantling of hoist or the finishing of concrete. See respective tables for these charges.

ABOVE GRADE WALLS

MANHOURS PER CUBIC YARD

Item	Manhours			
	Laborer	Oper. Engr.	Oiler	Total
Walls, ground to 8' high				
Hoist & buggies	.88	.16	—	1.04
Crane & bucket	.62	.07	.07	.76
Crane, bucket & buggies	.68	.09	.09	.86
Conveyor & buggies	.59	.07	—	.66
Walls, 8' to 16' high				
Hoist & buggies	1.17	.16	—	1.33
Crane & bucket	.83	.07	.07	.97
Crane, bucket & buggies	.91	.09	.09	1.09
Walls, 16' to 20' high				
Hoist & buggies	1.46	.16	—	1.62
Crane & bucket	1.04	.07	.07	1.18
Crane, bucket & buggies	1.14	.09	.09	1.32

Manhours are for the pouring and vibration of concrete for the above described walls in accordance with the introduction to this section.

Manhours do not include the fabrication, erection or dismantling of hoist or the finishing of concrete. See respective tables for these charges.

TILT-UP CONCRETE PANELS

MANHOURS PER SQUARE FOOT

Item	Manhours			
	Laborer	Cement Finisher	Iron Worker	Total
Pre-Cast Panels 6" Thick				
Place Concrete	.147	—	—	.147
Finish & Cure	.002	.015	—	.017
Erect Panels	—	—	.120	.120

Above manhours include the placement and vibration of concrete and the finishing, curing and erection of panels.

Manhours are based on the pouring of concrete in pre-erected forms on a flat level location easily accessible for both pouring and erection operations.

Placing concrete manhours are based on pouring concrete direct from ready-mix truck into forms.

Manhours do not include the fabrication, erection or stripping of forms. See respective tables for these charges.

RADIAL & RETAINING WALLS

MANHOURS PER CUBIC YARD

Item	Manhours				
	Laborer	Carpenter	Oper. Engr.	Oiler	Total
Radial Walls to 4' High					
Chute	.55	.04	–	–	.59
Crane & bucket	.65	–	.07	.07	.79
Radial Walls 4' to 12' High					
Crane & bucket	.85	–	.09	.09	1.03
Conveyor	.70	–	.07	–	.77
Retaining Wall to 10' High					
Crane & bucket	1.10	–	.10	.10	1.30
Conveyor	.90	–	.09	–	.99

Manhours include set-up time and the placement and vibration of concrete for the items as outlined above, in accordance with the introduction to this section.

Manhours do not include craft stand-by time, the fabrication of chutes or the finishing of concrete. See respective tables for these charges.

FLOOR SLABS, BEAMS & GIRDERS

MANHOURS PER CUBIC YARD

Item	Manhours				
	Laborer	Carpenter	Oper. Engr.	Oiler	Total
Square Beams & Girder					
Hoist & buggies	1.88	—	.19	—	2.07
Crane, bucket & buggies	1.43	—	.11	.11	1.65
Elevated Floor Slabs					
Hoist & buggies	1.56	—	.16	—	1.72
Crane, bucket & buggies	1.09	—	.10	.10	1.29
Ground Floor Slab					
Chute	.41	.04	—	—	.45
Buggies	.87	—	—	—	.87
Crane & bucket	.38	—	.08	.08	.54

Manhours include set-up time and placement and vibration of concrete as outlined above and in accordance with the introduction to this section.

Manhours do not include hoist or chute fabrication or hoist erection or finishing of concrete. See respective tables for these charges.

METAL PAN & T C TILE & RIB FLOORS

MANHOURS PER CUBIC YARD

Item	Manhours			
	Laborer	Oper. Engr.	Oiler	Total
Metal Pan Construction				
Hoist & buggies	1.76	.16	—	1.92
Crane, bucket & buggies	1.28	.10	.10	1.48
T C Tile and Rib Floors				
Hoist & buggies	1.81	.16	—	1.97
Crane, bucket & buggies	1.31	.10	.10	1.51

Above manhours include necessary set-up, placement and vibration or puddling of concrete, in accordance with the introduction to this section.

Manhours do not include hoist fabrication or installation, necessary craft stand-by time or concrete finishing. See respective tables for these charges.

PLACING LIGHTWEIGHT AND INSULATING CONCRETE AND FOAM FILLS ON FLOORS AND DECKS

MANHOURS PER CUBIC YARD

Item	Manhours			
	Laborer	Operator Engineer	Oiler	Total
On Ground Slabs				
Chute				
2″ thick	0.542	0.128	0.128	0.798
2½″ thick	0.614	0.145	0.145	0.904
3″ thick	0.693	0.163	0.163	1.019
3½″ thick	0.785	0.185	0.185	1.155
4″ thick	1.024	0.241	0.241	1.506
Buggies				
2″ thick	0.813	0.192	0.192	1.197
2½″ thick	0.921	0.218	0.218	1.357
3″ thick	1.040	0.245	0.245	1.530
3½″ thick	1.178	0.278	0.278	1.734
4″ thick	1.434	0.337	0.337	2.108
Crane & Bucket				
2″ thick	0.650	0.154	0.154	0.958
2½″ thick	0.737	0.174	0.174	1.085
3″ thick	0.832	0.196	0.196	1.224
3½″ thick	0.942	0.222	0.222	1.386
4″ thick	1.229	0.289	0.289	1.807
On Elevated Slabs				
Hoist & Buggies				
2″ thick	1.626	0.384	0.384	2.394
2½″ thick	1.842	0.435	0.435	2.712
3″ thick	2.079	0.489	0.489	3.057
3½″ thick	2.355	0.555	0.555	3.465
4″ thick	3.072	0.723	0.723	4.518
Crane, Bucket & Buggies				
2″ thick	1.219	0.288	0.288	1.795
2½″ thick	1.382	0.326	0.326	2.034
3″ thick	1.559	0.367	0.367	2.293
3½″ thick	1.766	0.416	0.416	2.598
4″ thick	2.304	0.542	0.542	3.388

Manhours are for the installation of items as outlined and are not to be confused with topping finish table on page 133.

Manhours do not include placing of forms or reinforcing, or curing and finishing. See respective tables for these time requirements.

RAMPS, PLATFORMS & CANOPIES

MANHOURS PER CUBIC YARD

Item	Manhours				
	Laborer	Carpenter	Oper. Engr.	Oiler	Total
Loading Dock & Ramp					
Chute	.60	.05	—	—	.65
Buggies	1.04	—	—	—	1.04
Crane & bucket	.80	—	.07	.07	.94
Conveyor & buggy	1.04	—	.10	—	1.05
Platforms					
Hoist & buggies	1.72	—	.18	—	1.90
Crane & bucket	1.09	—	.10	.10	1.29
Crane, bucket & buggies	1.20	—	.11	.11	1.42
Canopy					
Hoist & buggies	1.69	—	.15	—	1.84
Crane, bucket & buggies	1.20	—	.10	.10	1.40

Manhours include the necessary set-up time, placement of chutes and placement and vibration of concrete for the items as outlined above, in accordance with the introduction to this section.

Manhours do not include the fabrication of chutes or hoist, the erection of hoist or the finishing of concrete. See respective tables for these charges.

SILLS, LINTELS & COPING

(Poured in Place)

MANHOURS PER CUBIC FOOT

Item	Manhours			
	Laborer	Cement Finisher	Oper. Engr.	Total
Concrete Sills	.07	.03	—	.10
Concrete Lintels	.08	.04	.01	.13
Concrete Coping	.08	.04	.01	.13

Manhours are for the pouring in place of the above items and include necessary set-up and placement and vibration of concrete, in accordance with the introduction to this section.

Manhours do not include any hoist or chute fabrication, the placement of prefabricated or pre-cast items or the finishing of concrete. See respective tables for these charges.

STAIRWELLS & STAIRS

MANHOURS PER CUBIC YARD

Item	Manhours			
	Laborer	Oper. Engr.	Oiler	Total
Stairwells				
Hoist & Buggies	1.83	.20	—	2.03
Crane & bucket	1.30	.09	.09	1.48
Crane, Bucket & Buggies	1.43	.11	.11	1.65
Straight Stairs				
Hoist & Buggies	2.21	.23	—	2.44
Crane & Bucket & Buggies	1.65	.11	.11	1.87

Manhours are for the placement and vibration of concrete for the above operations, in accordance with the introduction to this section.

Manhours do not include the fabrication and erection of hoist or the finishing of concrete. See respective tables for these charges.

TRENCH & CURB CONCRETE

MANHOURS PER CUBIC YARD

Item	Manhours				
	Laborer	Carpenter	Oper. Engr.	Oiler	Total
Trench Walls & Slab					
Chute	.84	.06	–	–	.90
Buggies	1.47	–	–	–	1.47
Crane & bucket	1.25	–	.09	.09	1.43
Ground Curbs					
Buggies & hand shovel	1.35	–	–	–	1.35
Elevated Curbs					
Crane, bucket & buggy	2.06	–	.19	.19	2.44

Manhours are based on the pouring and vibration of concrete for the above listed items, in accordance with the introduction to this section.

Manhours do not include the fabrication of chutes or the finishing of concrete. See respective tables for these charges.

EQUIPMENT FOUNDATIONS

MANHOURS PER CUBIC YARD

Item	Manhours			
	Laborer	Oper. Engr.	Oiler	Total
Square Pads				
Crane & bucket	1.50	.19	.19	1.88
Crane, bucket & buggies	2.00	.25	.25	2.50
Offset, Skewed & Angled				
Crane & bucket	2.44	.38	.38	3.20
Crane, bucket & buggies	3.25	.50	.50	4.25

Manhours are for the placement and vibration of concrete for the above items, in accordance with the introduction to this section.

Square Pad manhours are based on pouring of square pads to four feet high either integral with floor or over pre-set dowels on-pre-poured floor.

Offset, skewed or angled manhours are based on that of pouring a large and bulky foundation with offsets or angles or both.

Manhours do not include finishing operations. See respective tables for these charges.

RUNWAYS & HOIST TOWERS

MANHOURS PER UNITS LISTED

Item	Unit	Manhours				
		Carpenter	Laborer	Iron Worker	Truck Driver	Total
Runways 3' to 5'						
Off Grade						
Fabricate, erect & remove	sq ft	.02	.02	—	.002	.042
Hoist Towers						
Wood - erect & remove	lin ft	1.85	1.51	—	.16	3.52
Tubular steel - erect & remove	lin ft	—	.65	1.89	.10	2.64

Manhours are for the fabrication, erection and removal of the above items and include handling and hauling of items from storage yard or warehouse to erection site within one thousand (1000) feet.

Manhours do not include the placement or finishing of concrete or concrete items. See respective tables for these charges.

TOPPING FINISH

MANHOURS PER SQUARE FOOT

Item	Manhours
	Cement Finisher
Integral Topping	
1/2" by hand	.029
1/2" by machine	.018
1" by hand	.068
1" by machine	.022
Separate Topping	
1/2" by hand	.036
1/2" by machine	.024
1" by hand	.072
1" by machine	.028
1-1/2" by hand	.080

Manhours are for topping finish and include mixing and stand-by time where required.

Manhours do not include the pouring and vibrating of main floors. See respective tables for these charges.

VAPOR BARRIER, CONCRETE FINISHING, GROUT, AND SANDBLASTING

MANHOURS PER UNITS LISTED

Item or Operation	Unit	Manhours
Vapor barrier—polyethylene—.004 to .010 mill thick	SF	.003
Screeding off slabs	SF	.006
Broom slabs	SF	.003
Wood float slabs	SF	.001
Hand steel trowels slabs	SF	.030
Machine trowel and hand burnish slabs	SF	.015
Edging or dividing	LF	.015
Acid clean and flush	SF	.013
Cure and protect	SF	.002
Remove fins or ties—point and patch	SF	.030
Carborundum rub	SF	.045
Exposed Aggregate Finish		
Bull float, machine float, machine trowel and wet brush	SF	.027
Stairways		
Treads only—hand trowel	SF	.023
Treads and risers—hand trowel	SF	.045
Panfilled stair treads	SF	.070
Abrasive non-slip	SF	.015
Grout column Bases—1″ Thick		
Portland cement	SF	.500
Non-shrink	SF	.575
Forms	LF	.030
Sandblast		
Light	SF	.025
Heavy	SF	.050

Manhours are for the above types of finish and include all necessary operations as may be required.

Manhours do not include the placement of concrete or concrete items. See respective tables for these charges.

In most areas, craft jurisdiction prevents laborers from helping cement finishers. Should this be the case, use total hours as listed above for cement finisher hours.

WATERPROOFING, DAMPPROOFING & INSULMASTIC

MANHOURS PER SQUARE FEET

Item	Manhours			
	Cement Finisher	Roofer	Insulator	Total
Foundation & Basement Walls				
Mastic type coating - 1/8"	.050	—	—	.050
Silicone or metallic coating	.020	—	—	.020
Damp resisting paint	.010	—	—	.010
Membrane & pitch (1 ply - 2 moppings)	—	.030	—	.030
Floors - Membrane & Pitch				
Dry floor (2 ply - 3 moppings)	—	.042	—	.042
Damp floor (2 ply - 3 moppings)	—	.067	—	.067
Application of Insulmastic				
Brushed on coat 1/16"	—	—	.020	.020
Brushed on coat 1/8"	—	—	.050	.050
Brushed on coat 1/4"	—	—	.068	.068

Manhours are for the waterproofing and dampproofing as itemized above and include all labor operations as may be necessary for this type of work.

Manhours do not include the placement of concrete or concrete items. See respective tables for these charges.

HARDENERS AND JOINT MATERIALS

MANHOURS PER UNITS LISTED

Item	Unit	Manhours			
		Cement Finisher	Carpenter	Laborer	Total
Metallic Grout Hardener					
Clean old surface	SF	–	–	.030	.030
One coat hardener	SF	.020	–	–	.020
Liquid Chemical Hardener					
Clean surface	SF	–	–	.030	.030
One coat hardener	SF	.010	–	–	.010
Joints					
Mastic compound	LF	.038	–	–	.038
Membrane course	LF	.018	–	–	.018
Asphalt Impregnated Expansion Joint					
Size to 1"x12"–in slab	LF	–	.020	–	.020
–in wall	LF	–	.035	–	.035
Rubber Water Stop					
Widths of 4", 5". & 6"–horizontal	LF	–	.031	–	.031
–vertical	LF	–	.042	–	.042
Widths of 9"–horizontal	LF	–	.042	–	.042
–vertical	LF	–	.054	–	.054
Metal Water Stop					
Widths of 4", 5". & 6"–horizontal	LF	–	.041	–	.041
–vertical	LF	–	.559	–	.559
Widths of 9"–horizontal	LF	–	.559	–	.559
–vertical	LF	–	.720	–	.720

Manhours are for the application or installation of the items and units listed and include all necessary labor operations as may be required for that particular item.

For asphalt expansion joint widths greater than 12 inches, use twenty-five (25) percent of the above manhours for each additional square foot or portion thereof.

Manhours do not include the placement of concrete or other concrete items. See respective tables for these time frames.

Section 10

MASONRY

It is not the intent of this section to cover all types and kinds of masonry work but rather to cover those which we are familiar with in some types of industrial construction.

The following manhour and percentage tables are based on averages of several projects installed under varied conditions where strict methods and preplanning were followed and strict reporting of actual cost was recorded.

We do not attempt to say that a mason can lay any given number of brick or tile in a day. Quite contrary to this, we have found that it is not what they can lay but what they will lay. This is dependent upon the conditions and location of the individual project. Conditions are covered under the introduction to this manual and should be thoroughly understood before attempting to apply the following manhour tables. As to location, we have actually found that productivity for this type of work is better in the southern and and southwestern states than in any other part of the country.

The following manhour tables cover items as may be required for this type of work in strict accordance with the preceding general notes and notes as listed on the individual manhour table pages.

GENERAL NOTES

The following notes apply to all masonry manhour tables except where noted on individual pages.

Manhours are based on continuous walls with normal amount of openings laid in common bond using full headers with ½ inch flush cut cement mortar joints and are average for interior and exterior work where applicable.

Mixing of mortar is assumed to be by machine. If mortar is to be handmixed, add 1.9 manhours of mortar mixer time per cubic yard mixed.

If type of bond or joint is to be different than those outlined above, apply percentages as appear on page 168 for these operations.

If lime mortar is to be used instead of cement mortar, decrease mason manhours in the following tables one (1) percent.

If walls are to be cavity type, increase total manhours in the tables by the following percentages:

Cavity walls to 10 inches thick	22 percent
Cavity walls to 20 inches thick	12 percent

Manhour tables include necessary time as may be required for the machine mixing and supplying of mortar to masons, the supplying and placing of masonry units in the desired location for installation by masons, and the placement or laying of masonry units by masons.

Hoisting engineer manhours are based on use of man-operated hoist. If automatic hoist is used to place units and mortar at higher level, eliminate hoist engineer manhours and add .25 Hod carrier or helper manhour per units as listed in the following tables.

Brick table manhours are listed for various heights in ten (10) feet sections up to and including fifty (50) feet in height. For heights not shown, always use manhours for next higher listing.

If walls are to be twelve (12) feet high, use manhours for twenty (20) feet height, etc. If walls are to be higher than fifty (50) feet, increase fifty (50) feet manhours by one-half of one (½ of 1) percent for each additional ten (10) feet or any part thereof, up to and including one hundred (100) feet and by one (1) percent for each additional ten (10) feet or any part thereof through two hundred (200) feet.

The following manhour tables do not include unloading and hauling of masonry units to installation site, the cleaning of masonry units after installation, caulking or pointing, installation of hoist or installation of scaffolding. The items are covered on individual tables and should be given separate consideration.

DOUBLE SIZE COMMON BRICK WALLS

MANHOURS PER THOUSAND (1000) BRICK

Item	Manhours			
	Mason	Hod Carrier	Hoist Engineer	Total
4" walls				
Ground to 5' elevation	21.94	17.94	–	39.88
Ground to 10' elevation	22.21	18.19	–	40.40
Ground to 20' elevation	22.35	18.29	–	40.64
Ground to 30' elevation	22.51	18.42	.07	41.00
Ground to 40' elevation	22.75	18.61	.07	41.43
Ground to 50' elevation	23.17	18.96	.08	42.21
8" Walls				
Ground to 5' elevation	20.29	15.95	–	36.24
Ground to 10' elevation	20.56	16.16	–	36.72
Ground to 20' elevation	20.69	16.26	–	36.95
Ground to 30' elevation	20.83	16.39	.07	37.29
Ground to 40' elevation	21.04	16.54	.07	37.65
Ground to 50' elevation	21.44	16.85	.08	38.37
12" Walls				
Ground to 5' elevation	17.47	15.50	–	32.97
Ground to 10' elevation	17.71	15.70	–	33.41
Ground to 20' elevation	17.82	15.79	–	33.61
Ground to 30' elevation	17.94	15.90	.07	33.91
Ground to 40' elevation	18.13	16.06	.07	34.26
Ground to 50' elevation	18.46	16.37	.08	34.91

Brick per square foot of wall:

4" walls	3.2 each
8" walls	6.4 each
12" walls	9.6 each

Cubic feet of mortar per 1000 brick:

4" wall — 13 cubic feet
8" wall — 19 cubic feet
12" wall — 22 cubic feet

COMMON BRICK WALLS

MANHOURS PER THOUSAND (1000) BRICK

Item	Manhours			
	Mason	Hod Carrier	Hoist Engineer	Total
4" or 4½" Walls				
Ground to 5' elevation	13.71	11.21	—	24.92
Ground to 10' elevation	13.88	11.37	—	25.25
Ground to 20' elevation	13.97	11.43	—	25.40
Ground to 30' elevation	14.07	11.51	.05	25.63
Ground to 40' elevation	14.22	11.63	.06	25.91
Ground to 50' elevation	14.48	11.85	.07	26.40
8" or 9" Walls				
Ground to 5' elevation	12.68	9.97	—	22.65
Ground to 10' elevation	12.85	10.10	—	22.95
Ground to 20' elevation	12.93	10.16	—	23.09
Ground to 30' elevation	13.02	10.23	.05	23.30
Ground to 40' elevation	13.15	10.34	.06	23.55
Ground to 50' elevation	13.40	10.53	.07	24.00
12" or 13" Walls				
Ground to 5' elevation	10.92	9.69	—	20.61
Ground to 10' elevation	11.07	9.81	—	20.88
Ground to 20' elevation	11.14	9.87	—	21.01
Ground to 30' elevation	11.21	9.94	.05	21.20
Ground to 40' elevation	11.33	10.04	.06	21.43
Ground to 50' elevation	11.54	10.23	.07	21.84

Brick per square foot of wall:

4" or 4½" walls	6.35 each
8" or 9" walls	12.70 each
12" or 13" walls	19.05 each

Approximately 18 cubic feet of mortar per 1000 brick.

COMMON BRICK WALLS

(CONTINUED)

MANHOURS PER THOUSAND (1000) BRICK

Item	Mason	Hod Carrier	Hoist Engineer	Total
16" or 17" Walls				
Ground to 5' elevation	10.70	9.50	—	20.20
Ground to 10' elevation	10.89	9.61	—	20.50
Ground to 20' elevation	10.92	9.67	—	20.59
Ground to 30' elevation	10.99	9.74	.05	20.78
Ground to 40' elevation	11.10	9.89	.06	21.05
Ground to 50' elevation	11.31	10.03	.07	21.41
20" or 21" Walls				
Ground to 5' elevation	10.38	9.22	—	19.60
Ground to 10' elevation	10.56	9.32	—	19.88
Ground to 20' elevation	10.59	9.38	—	19.97
Ground to 30' elevation	10.66	9.45	.05	20.16
Ground to 40' elevation	10.78	9.59	.06	20.43
Ground to 50' elevation	10.97	9.73	.07	20.77
24" to 25" Walls				
Ground to 5' elevation	9.96	8.85	—	18.81
Ground to 10' elevation	10.14	8.95	—	19.09
Ground to 20' elevation	10.17	9.00	—	19.17
Ground to 30' elevation	10.23	9.07	.05	19.35
Ground to 40' elevation	10.35	9.21	.06	19.62
Ground to 50" elevation	10.53	9.34	.07	19.94

Brick per square foot of wall:

16" or 17" walls	25.40 each
20" or 21" walls	31.75 each
24" or 25" walls	38.10 each

Approximately 18 cubic feet of mortar per 1000 brick.

FACE BRICK WITH COMMON BRICK BACK-UP WALLS

MANHOURS PER THOUSAND (1000) BRICK

Item	Manhours			
	Mason	Hod Carrier	Hoist Engineer	Total
4" Face Brick Wall				
Ground to 5' elevation	20.57	11.21	—	31.78
Ground to 10' elevation	20.82	11.37	—	32.19
Ground to 20' elevation	20.96	11.43	—	32.39
Ground to 30' elevation	21.11	11.51	.05	32.67
Ground to 40' elevation	21.33	11.63	.06	33.02
Ground to 50' elevation	21.72	11.85	.07	33.64
8" Wall — 4" Face & 4" Common Brick				
Ground to 5' elevation	19.02	9.97	—	28.99
Ground to 10' elevation	19.28	10.10	—	29.38
Ground to 20' elevation	19.40	10.16	—	29.56
Ground to 30' elevation	19.53	10.23	.05	29.81
Ground to 40' elevation	19.73	10.34	.06	30.13
Ground to 50' elevation	19.91	10.53	.07	30.51
12" Wall — 4" Face & 8" Common Brick				
Ground to 5' elevation	16.38	9.69	—	26.07
Ground to 10' elevation	16.61	9.81	—	26.42
Ground to 20' elevation	16.71	9.87	—	26.58
Ground to 30' elevation	16.82	9.94	.05	26.81
Ground to 40' elevation	17.00	10.04	.06	27.10
Ground to 50' elevation	17.31	10.23	.07	27.61

Brick per square foot of wall:

4" walls	6.35 each
8" walls	12.70 each
12" walls	19.05 each

Approximately 18 cubic feet of mortar per 1000 brick.

FACE BRICK WITH COMMON BRICK BACK-UP WALLS

(CONTINUED)

MANHOURS PER THOUSAND (1000) BRICK

Item	Mason	Hod Carrier	Hoist Engineer	Total
16" Wall — 4" Face & 12"				
Common Brick				
Ground to 5' elevation	16.05	9.50	—	25.55
Ground to 10' elevation	16.34	9.61	—	25.95
Ground to 20' elevation	16.38	9.67	—	26.05
Ground to 30' elevation	16.49	9.74	.05	26.28
Ground to 40' elevation	16.65	9.89	.06	26.60
Ground to 50' elevation	16.97	10.03	.07	27.07
20" Wall — 4" Face & 16"				
Common Brick				
Ground to 5' elevation	15.57	9.22	—	24.79
Ground to 10' elevation	15.84	9.32	—	25.16
Ground to 20' elevation	15.89	9.38	—	25.27
Ground to 30' elevation	15.99	9.45	.05	25.49
Ground to 40' elevation	16.17	9.59	.06	25.82
Ground to 50' elevation	16.46	9.73	.07	26.26
24" Wall — 4" Face & 20"				
Common Brick				
Ground to 5' elevation	14.94	8.85	—	23.79
Ground to 10' elevation	15.21	8.95	—	24.16
Ground to 20' elevation	15.26	9.00	—	24.26
Ground to 30" elevation	15.35	9.07	.03	24.47
Ground to 40' elevation	15.53	9.21	.06	24.80
Ground to 50' elevation	15.80	9.34	.07	25.21

Brick per square foot of walls:

16" walls	25.40 each
20" walls	31.75 each
24" walls	38.10 each

Approximately 18 cubic feet of mortar per 1000 brick.

MODULAR BRICK

MANHOURS PER THOUSAND (1000) BRICK

Item	Manhours			
	Mason	Hod Carrier	Hoist Engineer	Total-
4" Walls				
Ground to 5' elevation	12.40	11.21	–	23.61
Ground to 10' elevation	12.49	11.37	–	23.86
Ground to 20' elevation	12.57	11.43	–	24.00
Ground to 30' elevation	12.66	11.51	.05	24.22
Ground to 40' elevation	12.80	11.63	.06	24.49
Ground to 50' elevation	13.03	11.85	.07	24.95
8" Walls				
Ground to 5' elevation	11.41	9.97	–	21.38
Ground to 10' elevation	11.57	10.10	–	21.67
Ground to 20' elevation	11.64	10.16	–	21.80
Ground to 30' elevation	11.72	10.23	.05	22.00
Ground to 40' elevation	11.84	10.34	.06	22.24
Ground to 50' elevation	12.06	10.53	.07	22.66
12" Walls				
Ground to 5' elevation	9.83	9.69	–	19.52
Ground to 10' elevation	9.96	9.81	–	19.77
Ground to 20' elevation	10.03	9.87	–	19.90
Ground to 30' elevation	10.09	9.94	.05	20.08
Ground to 40' elevation	10.20	10.04	.06	20.30
Ground to 50' elevation	10.39	10.23	.07	20.69

Number of brick per square foot:

Size	1-brick	2-brick	3-brick
2-1/6" x 3-1/2" x 7-1/2"	6.75	13.50	20.25
2-1/4" x 3-5/8" x 7-5/8"	6.75	13.50	20.25
2-5/16" x 3-5/8" x 7-5/8"	6.75	13.50	20.25
2-1/2" x 3-1/2" x 7-1/2"	6.00	12.00	18.00
3-1/2" x 3-1/2" x 7-1/2"	4.50	9.00	13.50

Mortar will vary according to size of units used. We have found 18 cubic feet per 1000 brick to be a good average.

BRICK WITH LOAD BEARING CLAY TILE BACK-UP

MANHOURS PER HUNDRED (100) SQUARE FEET

Item	Manhours			
	Mason	Hod Carrier	Hoist Engr.	Total
8" Wall Common brick with 3-3/4" x 12" x 12" tile	12.31	10.08	.05	22.44
Face brick with 3-3/4" x 12" x 12" tile	14.37	11.76	.05	26.18
12" Wall Common brick with 8" x 12" x 12" tile	14.77	12.09	.05	26.91
Face brick with 8" x 12" x 12" tile	16.83	13.77	.05	30.65

Manhours are average for heights to fifty (50) feet.

Brick and Tile Per Hundred Square Feet of Wall:

Wall Size	Brick Each	Tile Each
8"	635	100
12"	635	100

Approximate Mortar required per 100 Square Feet of Wall:

8" wall	14 cubic feet
12" wall	17 cubic feet

BRICK WITH CONCRETE BLOCK BACK-UP

MANHOURS PER HUNDRED (100) SQUARE FEET

Item	Manhours			
	Mason	Hod Carrier	Hoist Engr.	Total
8" Wall				
Common brick with 3-5/8" x 7-5/8" x 15-5/8" block	13.56	11.09	.06	24.71
Face brick with 3-5/8" x 7-5/8" x 15-5/8" block	15.61	12.78	.06	28.45
12" Wall				
Common brick with 7-5/8" x 7-5/8" x 15-5/8" block	15.52	12.70	.06	28.28
Face brick with 7-5/8" x 7-5/8" x 15-5/8" block	17.58	14.38	.06	32.02

Manhours are average for heights to fifty (50) feet.

Brick and Concrete Block per hundred square feet of wall:

Wall Size	Brick Each	Conc. Blk. Ea.
8"	635	112.5
12"	635	112.5

Approximate Mortar required per hundred square feet of wall:

8" wall	17 cubic feet
12" wall	18 cubic feet

CONCRETE BRICK WALLS

MANHOURS PER THOUSAND (1000) BRICK

Item	Manhours			
	Mason	Hod Carrier	Hoist Engr.	Total
8" Wall				
Ground to 5' elevation	15.85	9.97	–	25.82
Ground to 10' elevation	16.06	10.10	–	26.16
Ground to 20' elevation	16.16	10.16	–	26.32
Ground to 30' elevation	16.28	10.23	.05	26.56
Ground to 40' elevation	16.44	10.34	.06	26.84
Ground to 50' elevation	16.75	10.53	.07	27.35
12" Wall				
Ground to 5' elevation	13.65	9.69	–	23.34
Ground to 10' elevation	13.84	9.81	–	23.65
Ground to 20' elevation	13.93	9.87	–	23.80
Ground to 30' elevation	14.01	9.94	.05	24.00
Ground to 40' elevation	14.16	10.04	.06	24.26
Ground to 50' elevation	14.43	10.23	.07	24.73

Above manhours are average for standard, modular and jumbo-size concrete brick.

Mortar Joints assumed at:
  Stand Brick - Vertical Joints 1/4", Horizontal Joints 1/2"
  Modular & Jumbo - Vertical & Horizontal Joints at 3/8"

Brick Per Square Foot of Wall:
  Standard (2-1/4" x 3-3/4" x 8") 8" wall, 14 each; 12" wall, 21 each.
  Modular (2-1/4" x 3-5/8" x 7-5/8") 8" wall, 13.5 each; 12" wall, 20.25 each.
  Jumbo (3-5/8" x 3-5/8" x 7-5/8") 8" wall, 9 each; 12" wall, 13.50 each.

LOAD BEARING CONCRETE BLOCK WALLS

MANHOURS PER HUNDRED (100) BLOCK

Block Size	No. Blocks Per 100 Sq. Ft. of Wall	Cu. Ft. Mortar Per 100 Sq. Ft. of Wall	Manhours			
			Mason	Hod Carrier	Hoist Engr.	Total
3-5/8" x 4-7/8" x 11-5/8"	225	8.0	3.7	3.5	.5	7.7
5-5/8" x 4-7/8" x 11-5/8"	225	8.5	4.2	3.9	.5	7.6
7-5/8" x 4-7/8" x 11-5/8"	225	9.0	4.5	4.3	.5	9.3
3-5/8" x 7-5/8" x 11-5/8"	150	6.0	4.2	3.9	.5	8.6
5-5/8" x 7-5/8" x 11-5/8"	150	6.5	4.5	4.3	.5	9.3
7-5/8" x 7-5/8" x 11-5/8"	150	7.0	5.1	4.8	.6	9.5
9-5/8" x 7-5/8" x 11-5/8"	150	7.5	5.7	5.5	.7	11.9
11-5/8" x 7-5/8" x 11-5/8"	150	8.0	6.7	6.4	.7	14.0
3-5/8" x 3-5/8" x 15-5/8"	225	9.0	3.7	3.5	.5	7.7
5-5/8" x 3-5/8" x 15-5/8"	225	9.5	4.2	3.9	.5	·8.6
7-5/8" x 3-5/8" x 15-5/8"	225	10.0	4.5	4.3	.5	9.3
3-5/8" x 7-5/8" x 15-5/8"	112.5	5.0	4.3	4.0	.5	8.8
5-5/8" x 7-5/8" x 15-5/8"	112.5	5.0	4.8	4.5	.5	9.8
7-5/8" x 7-5/8" x 15-5/8"	112.5	6.0	5.3	5.0	.6	10.9
9-5/8" x 7-5/8" x 15-5/8"	112.5	6.5	6.0	5.7	.7	12.4
11-5/8" x 7-5/8" x 15-5/8"	112.5	7.0	7.0	6.7	.7	14.4

Manhours are based on average of continuous walls through 24 feet in height. If walls are to be higher than 24 feet, increase above manhours by one (1) percent for each additional 5 feet or any part thereof.

For walls below grade, decrease manhours by 10 percent.

For light-weight concrete block walls above grade, decrease above manhours 5 percent.

Increase 7⅝-inch high block manhours fifteen (15) percent for installation of split face units. twenty (20) percent for split face "Fluted" units, and ten (10) percent for slump blocks.

For garden or fence walls, decrease manhours fifteen (15) percent.

Where one side of wall is to be covered or concealed from view reduce above manhours ten (10) percent.

For standard block cut allow 0.10 manhour each.

Allow 0.10 manhour per linear foot measured around door opening for shoring of opening.

If use of hoist operated by hoist operator is not necessary, eliminate above hoist engineer manhours.

SPECTRA-GLAZE AND SCREEN WALL BLOCK

MANHOURS PER HUNDRED (100) BLOCK

Block size and Type	No. Blocks Per 100 Sq Ft of Wall	Cu Ft Mortar Per 100 Sq Ft of Wall	Manhours			
			Mason	Hod Carrier	Hoist Engineer	Total
Spectra-Glaze Blocks						
Glazed One Side						
3⅝"x7⅝"x15⅝"	112.5	5.0	7.5	7.0	0.8	15.3
5⅝"x7⅝"x15⅝"	112.5	5.0	8.2	7.7	0.9	16.8
7⅝"x7⅝"x15⅝"	112.5	6.0	8.9	8.4	1.0	18.3
11⅝"x7⅝"x15⅝"	112.5	7.0	11.6	11.1	1.2	22.9
Glazed Two Sides						
3⅝"x7⅝"x15⅝"	112.5	5.0	8.3	7.7	0.8	16.8
5⅝"x7⅝"x15⅝"	112.5	5.0	9.0	8.4	0.9	18.3
7⅝"x7⅝"x15⅝"	112.5	6.0	9.8	9.3	1.0	20.1
Screen Wall Block						
3⅝"x7⅝"x11⅝"	112.5	6.0	3.6	3.4	–	7.0
3⅝"x11⅝"x11⅝"	100.0	8.0	4.5	4.3	–	8.8

Spectra-Glaze manhours are based on average of continuous walls through 16 feet in height. If walls are to be higher than 16 feet, increase manhours by one (1) percent for each additional 5 feet or any part thereof.

If use of hoist, operated by hoist operator is not necessary, eliminate hoist engineer manhours.

FIRE BRICK & TILE

MANHOURS PER UNITS LISTED

Item	Unit	Manhours			
		Mason	Hod Carrier	Hoist Engineer	Total
Lining Chimneys & stacks with Fire Brick	1,000	17.0	12.0	.05	29.5
Fire Brick in Fire-Boxes, Breechings, Etc.	1,000	42.0	38.0	—	80.0
Bricking Boilers	1,000	14.0	12.0	—	26.0
Clay Tile Flue Lining	sq ft	.1	.1	—	0.2
Radial Stack (125' high, 13' diameter at bottom and 6' at top lined 60'	Ton	3.0	5.0	.9	8.9

Lining chimney and stack units: Manhours are average for small stacks, for this type of work laid in fire clay.

Fire brick in fire-boxes, breechings, etc., units: Average manhours for placement of arch, radial brick and special shapes.

Bricking boiler units: Average manhours for bricking in steel boilers, fire-boxes and breeching with common or fire brick where the walls are 8" to 12" thick.

Clay time flue lining units: Average manhours per square foot of lining flues. This item is usually estimated by the linear foot by size. The above manhours are average of many jobs of various shapes and sizes and have been changed to square feet for convenience of the estimator.

Radial stack units: Average manhours of three similar type stacks as described above and are per ton of brick in place.

Manhours do not include installation of hoist or jib or scaffolding. See respective tables for these charges.

HOLLOW CLAY PARTITION & FURRING TILE

MANHOURS PER HUNDRED (100) SQUARE FEET

Tile Size	Manhours			
	Mason	Hod Carrier	Hoist Engr.	Total
2" x 12" x 12" (split furring)	4.6	4.6	.3	9.5
2" x 12" x 12"	3.8	3.8	.3	7.9
3" x 12" x 12"	3.8	3.8	.3	7.9
4" x 12" x 12"	4.2	4.2	.3	8.7
5" x 12" x 12"	4.3	4.3	.4	9.0
6" x 12" x 12"	4.8	4.8	.4	10.0
8" x 12" x 12"	5.8	5.8	.5	12.1
10" x 12" x 12"	7.2	7.2	.5	14.9
12" x 12" x 12"	7.7	7.7	.5	15.9

Manhours are based on partition heights of 8'0" for 2" tile, 12'0" for 3" tile, 16'0" for 4" tile and 20'0" to 24'0" for all other size tiles.

Manhours are based on square foot of single tile. If combination of two tiles are used to fill out partition, double above manhours.

If strips of wire mesh reinforcing is added between each course of tile, increase above manhours eight (8) percent.

All above tile is equal to one (1) square foot each.

Cubic feet of mortar per 100 square feet of tile using 1/2" bed and end joint:

Cells Laid Horizontal
2" x 12" x 12" — 1.50 cu ft
3" x 12" x 12" — 1.60 cu ft
4" x 12" x 12" — 2.00 cu ft
6" x 12" x 12" — 2.50 cu ft
8" x 12" x 12" — 3.10 cu ft
10" x 12" x 12" — 4.00 cu ft
12" x 12" x 12" — 5.00 cu ft

Cells Laid Vertical
6" x 12" x 12" — 3.50 cu ft
8" x 12" x 12" — 4.50 cu ft
10" x 12" x 12" — 5.30 cu ft
12" x 12" x 12" — 6.00 cu ft

GYPSUM TILE

MANHOURS PER HUNDRED (100) SQUARE FEET

Tile Size	Manhours			
	Mason	Hod Carrier	Hoist Engr.	Total
2" x 12" x 30" - Solid	2.8	2.9	.3	6.0
3" x 12" x 30" - Hollow	3.0	3.1	.3	6.4
4" x 12" x 30" - Hollow	3.4	3.5	.3	7.2
6" x 12" x 30" - Hollow	3.8	3.9	.5	8.2

Manhours are based on partition heights of 10'0" for 2" tile, 13'0" for 3" tile, 17'0" for 4" tile and 30'0" for 6" tile.

All above size tile are equal to 2.5 square feet each.

Assuming 1/2 inch bed and end mortar joint of gypsum partition tile cement, the following mortar quantity should install 100 square feet of tile:

2" Tile — 2 cu ft
3" Tile — 2½ cu ft
4" Tile — 3 cu ft
6" Tile — 4 cu ft

LOAD BEARING TILE WALLS

MANHOURS PER HUNDRED (100) SQUARE FEET

Tile Size	Manhours			
	Mason	Hod Carrier	Hoist Engr.	Total
3¾" x 12" x 12"	4.3	4.3	.4	9.0
6" x 12" x 12"	5.3	5.3	.4	11.0
8" x 12" x 12"	6.6	6.6	.5	13.7
10" x 12" x 12"	7.4	7.4	.6	15.4
12" x 12" x 12"	8.2	8.2	.6	17.0

Manhours are average for heights to 30'0", using ¼" mortar joints.

Any of the above tile equals one (1) square foot in area.

Cubic feet of mortar required per 100 square feet of tile:

3¾" x 12" x 12"	2.00 cubic feet
6" x 12" x 12"	3.30 cubic feet
8" x 12" x 12"	4.60 cubic feet
10" x 12" x 12"	5.30 cubic feet
12" x 12" x 12"	5.90 cubic feet

GLAZED ONE SIDE STRUCTURAL FACING TILE

MANHOURS PER HUNDRED (100) PIECES

Nominal Tile Size	Manhours			
	Mason	Hod Carrier	Hoist Engr.	Total
2" x 5-1/3" x 8"	5.0	5.0	.3	10.3
2" x 4" x 12"	6.8	6.8	.3	13.9
2" x 5-1/3" x 12"	6.8	6.8	.3	13.9
2" x 8" x 16"	10.0	10.0	.3	20.3
4" x 5-1/3" x 8"	5.1	5.1	.5	10.7
4" x 4" x 12"	6.9	6.9	.5	14.3
4" x 5-1/3" x 12"	6.9	6.9	.5	14.3
4" x 8" x 16"	13.5	13.5	.5	27.5
6" x 5-1/3" x 8"	5.2	5.2	.6	11.0
6" x 4" x 12"	6.6	6.6	.6	13.8
6" x 5-1/3" x 12"	6.8	6.8	.6	14.2
8" x 5-1/3" x 8"	6.0	6.0	.7	12.7
8" x 4" x 12"	7.2	7.2	.7	15.1
8" x 5-1/3" x 12"	8.9	8.9	.7	18.5

Manhours are average of several jobs of various heights using the above size soaps and tile laid with ¼ inch mortar joints.

If hoisting engineer is not required, eliminate the above hoisting engineer manhours.

Manhours do not include plastering of unglazed side. See respective table for this charge.

GLAZED TWO SIDES STRUCTURAL FACING TILE

MANHOURS PER HUNDRED (100) PIECES

Nominal Tile Size	Manhours			
	Mason	Hod Carrier	Hoist Engr.	Total
4" x 5-1/3" x 8"	5.9	5.9	.5	12.3
4" x 4" x 12"	7.0	7.0	.5	14.5
4" x 5-1/3" x 12"	7.0	7.0	.5	14.5
6" x 5-1/3" x 8"	6.0	6.0	.6	12.6
6" x 4" x 12"	7.2	7.2	.6	15.0
6" x 5-1/3" x 12"	7.2	7.2	.6	15.0
8" x 5-1/3" x 8"	6.0	6.0	.7	12.7
8" x 4" x 12"	7.3	7.3	.7	15.3
8" x 5-1/3" x 12"	7.3	7.3	.7	15.3

Manhours are average for glazed two-sides tile of several jobs using the above size tile laid in ¼ inch mortar joints.

If hoisting engineer is not required, eliminate the above hoisting engineer manhours.

GLASS BLOCK & ACCESSORIES

MANHOURS PER HUNDRED (100) SQUARE FEET

Item	Manhours		
	Mason	Hod Carrier	Total
Glass Block:			
5-3/4" x 5-3/4" x 3-7/8"	19.	11.	30.
7-3/4" x 7-3/4" x 3-7/8"	14.	8.	22.
11-3/4" x 11-3/4" x 3-7/8"	10.	8.	18.
Ramming Oakum & caulking	2.	—	2.
Cleaning Blocks	3.	—	3.

Manhours are based on 10'0" x 10'0" panels and include necessary operations for the type work as outlined above.

Mortar to be stiff lime type with joints 3/16" to 3/8" thick.   Mortar required per 100 square feet as is follows:

Block	5-3/4" x 5-3/4" x 3-7/8"	13 cubic feet
Block	7-3/4" x 7-3/4" x 3-7/8"	16 cubic feet
Block	11-3/4" x 11-3/4" x 3-7/8"	24 cubic feet

Above block manhours include concave tooled mortar joints after initial set.

Number of blocks required per square feet of wall:

5-3/4" x 5-3/4" x 3-7/8"	4.00 each
7-3/4" x 7-3/4" x 3-7/8"	2.25 each
11-3/4" x 11-3/4" x 3-7/8"	1.00 each

ARCHITECTURAL TERRA COTTA

MANHOURS PER HUNDRED (100) CUBIC FEET

Item	Manhours			
	Mason	Hod Carrier	Oper. Engr.	Total
Hoist & Place				
Large pieces	12.00	20.00	3.20	35.20
Complicated or intricate	27.00	50.00	4.50	81.50
Trim	18.40	29.60	4.50	52.50
Point & Clean	8.40	7.20	—	15.60

Above manhours are average of several jobs and include operations as may be necessary for this type of work.

Mortar requirements are based on ¼ cubic yard per 100 cubic feet of architectural terra cotta work using a lime putty type.

Manhours do not include scaffolding. See respective table for this charge.

ASHLAR VENEER STONE & GRANITE FACING

MANHOURS PER HUNDRED (100) SQUARE FEET

Item	Manhours			
	Mason	Hod Carrier	Oper. Engr.	Total
Ashlar Veneer Stone 4" or 6"				
Hoist & place	9.60	16.00	2.40	26.00
Point & clean	4.00	4.00	–	8.00
Granite Facing				
Walls	17.60	12.80	–	30.40
Columns	27.20	16.80	–	44.00
Wainscot	16.00	11.20	–	27.20

Above manhours are average of several jobs and include operations as may be necessary for this type of work.

Manhours include unloading, handling, storing and placing.

Manhours do not include scaffolding. See respective table for this charge.

SILLS, LINTELS & COPINGS

MANHOURS PER HUNDRED (100) LINEAR FEET

Item	Manhours			
	Mason	Hod Carrier	Hoist Engr.	Total
Sills				
Limestone (5" x 8")	7.40	12.60	2.30	22.30
Brick	4.00	4.00	—	8.00
Precast concrete	5.90	11.40	2.40	19.70
Lintels				
Limestone	7.50	12.80	2.50	22.80
Precast concrete	6.00	11.50	2.40	19.90
Copings				
Limestone (1-1/3" x 5")	7.20	12.75	2.40	22.35
Precast concrete	5.80	11.25	2.40	19.45
Tile	4.67	7.46	2.00	14.13

Manhours are average of many jobs of this nature and include handling, hauling and placing in proper mortar bed.

Manhours do not include precasting or scaffolding. For scaffolding see respective table.

MISCELLANEOUS INTERIOR MARBLE WORK

MANHOURS PER UNITS LISTED

Item	Unit	Manhours				
		Mason	Hod Carrier	Laborer	Hoist Engr.	Total
Marble Ashlar Wall Facing	100 lin ft	12.00	12.00	9.00	1.00	34.00
Marble Wainscot	100 sq ft	8.00	15.00	–	–	23.00
Marble Wainscot Cap	100 lin ft	9.60	15.00	–	–	24.60
Marble Column Facing	100 sq ft	4.00	21.00	–	21.00	46.00
Marble Door & Window Trim	100 lin ft	6.00	13.50	–	13.00	31.50
Marble Toilet Stalls						
Partitions	100 sq ft	12.80	9.00	6.40	–	28.20
Stiles	100 lin ft	14.80	13.88	6.40	–	35.08
Cap	100 lin ft	11.20	9.00	6.40	–	26.60

Manhours are average of several jobs of this nature and include handling, hauling, job fabrication and placing.

Manhours do not include scaffolding. See respective table for this charge.

CLOSE UP OPENINGS WITH BRICK OR TILE

MANHOURS PER SQUARE FOOT

Item	Manhours		
	Mason	Hod Carrier	Total
Openings to 25 Square Feet in Area			
4" Brick	.15	.08	.23
4" Tile	.08	.04	.12
8" Brick	.20	.12	.32
8" Tile	.12	.08	.20
12" Brick	.22	.12	.34
Openings 25 to 50 Square Feet in Area			
4" Brick	.12	.07	.19
4" Tile	.06	.04	.10
8" Brick	.12	.09	.21
8" Tile	.08	.06	.14
12" Brick	.15	.10	.25
Openings 50 to 100 Square Feet in Area			
4" Brick	.10	.08	.18
4" Tile	.05	.03	.08
8" Brick	.11	.09	.20
8" Tile	.07	.05	.12
12" Brick	.13	.11	.24

Manhours are average for closing openings in old walls of the sizes as outlined above with brick or tile using cement mortar.

Manhours do not include scaffolding. See respective table for this charge.

WALL REINFORCING AND TIES

MANHOURS PER UNITS LISTED

Item	Unit	Manhours Mason	Manhours Helper	Manhours Total
Reinforcing Bars				
Block Walls				
Placed horizontally	C LF	0.49	0.21	.070
Placed vertically	C LF	0.63	0.27	.090
Bond Beams				
Placed horizontally	C LF	0.63	0.27	0.90
Placed vertically	C LF	0.78	0.34	1.12
Brick Walls				
Placed horizontally	C LF	0.78	0.34	1.12
Placed vertically	C LF	0.95	0.40	1.35
Truss or Tri-Rod Reinforcing				
Block Walls				
Placed horizontally	C LF	0.38	0.16	0.54
Brick Walls				
Placed horizontally	C LF	0.38	0.16	0.54
Metal Wall Ties				
3/16″ Rectangular Ties				
Regular—4″x6″ or 4″x8″	C EA	0.38	0.16	0.54
Adjustable—4″x6″. 4″x8″ or 4″x10″	C EA	0.50	0.22	0.72
3/16″ "Z" Type Brick Ties				
Regular—6″ or 8″ long	C EA	0.38	0.16	0.54
Adjustable—6″ or 8″ long	C EA	0.50	0.22	0.72

Reinforcing bar manhours are only for placing of pre-cut. pre-fabricated number 3. 4. 5. and 6 bars.

Truss of tri-rod reinforcing manhours are only for placing of number 3. 4. 5. 6. 8. 10. and 12 sizes. Ladder type truss reinforcing can also be placed for these manhours.

Metal wall tie manhours are for the placement of ties as outlined.

Union rules may require the use of iron workers for placement of reinforcing. If this is the case, substitute iron worker time for mason time listed above.

GROUTING AND CAULKING MASONRY WALLS

MANHOURS PER UNITS LISTED

Item		Manhours			
	Unit	Mason	Helper	Hoist Engineer	Total
Concrete Block Walls					
Ground below grade single standard block	CY	1.20	2.10	–	3.30
Grout below grade bond beam block	CY	1.25	2.15	–	3.40
Grout above grade bond beam block	CY	1.05	2.05	0.50	3.60
Grout above grade lintel bond beam block	CY	0.90	1.80	0.45	3.15
Grout above grade vertical cells of block	CY	0.90	1.80	0.45	3.15
Brick Walls					
Solid grout cavities	CY	0.80	1.55	0.35	2.70
Fill vertical pilasters	CY	0.90	1.80	0.45	3.15
Grout fill horizontal beams and lintels	CY	1.05	2.05	0.50	3.60
Fill and Caulk Joints					
Place polyurethane foam and caulking	LF	0.05	–	–	0.05

Grouting manhours include mixing, hoisting when necessary and placing of grout for the various items as listed.

Helper manhours include grout mixing hours.

Fill and caulk joint manhours include placement of foam filler and caulking compound.

For cast in place beams and lintels see Formworks, Reinforcing, and Concrete Sections.

BOLTS, FLASHINGS, FILLS, AND INSULATIONS

MANHOURS PER UNITS LISTED

Item Description	Unit	Manhours		
		Mason	Helper	Total
Anchor Bolts				
Straight type through size ½"x18"	EA	.067	.034	.101
"J" type through size ¾"x12"	EA	.135	.067	.202
Thru Wall Flashing				
30# asphalt felt	SF	.034	.017	.051
Galvanized steel 26 through 20 gauge	SF	.067	.034	.101
Aluminum 0.020 and 0.025	SF	.050	.034	.084
30 mil polyvinyl chloride	SF	.034	.017	.051
Copper covered polyethylene	SF	.050	.034	.084
30# felt underlayment	SF	.034	.017	.051
Mortar fill door frames	CF	.050	.100	.150
Set lintel angles	LF	.150	.100	.250
Reglet—Extruded aluminum 0.55"	LF	.050	.034	.084
Insulation				
Zonolite—Cavity walls	CF	.034	.017	.051
Zonolite—Concrete block walls	CF	.050	.034	.084
1" thick polystyrene applied with glue	SF	.034	.017	.051

Anchor bolt manhours are for handling and setting bolts without templates. If templates are required or used, the manhours should be doubled.

Flashing or underlayment manhours are for handling and placing flashings or underlayment in or on walls.

Mortar fill door frame manhours are for mixing and filling voids in door frames with mortar.

Set lintel manhours include handling and placing angle lintels. 3"x3"x3/16" through 4"x4"x½", in walls.

Reglet manhours include handling and placing reglet.

Insulation manhours include handling and placing of insulation of the types listed.

UNLOADING MASONRY MATERIALS

MANHOURS PER THOUSAND

Item	Manhours		
	Laborer	Truck Driver	Total
Brick			
Face	2.90	1.85	4.80
Common	1.25	1.85	3.10
Jumbo or Utility	3.75	1.85	5.60
Partition & Load Bearing Clay Tile			
4" Tile	7.80	1.85	9.65
6" Tile	9.90	1.95	11.85
8" Tile	11.75	2.20	13.95
10" Tile	12.50	2.40	14.90
12" Tile	13.00	2.60	15.60
Glazed Tile			
8" Tile (Length)	10.50	2.40	12.90
12" Tile (Length)	12.75	2.80	15.55
16" Tile (Length)	15.00	3.00	18.00

Manhours include unloading, handling, hauling to storage or erection site within 1000 feet and stacking.

Manhours do not include installation. See respective tables for this charge.

POINT & CAULK AROUND SASH & CLEAN WALLS

MANHOURS PER UNITS LISTED

Item	Unit	Manhours		
		Mason	Hod Carrier	Total
Point & Caulk Around Sash				
Openings				
Point with mortar	100 lin ft	2.0	1.0	3.0
Caulk with caulking compound	100 lin ft	1.6	.8	2.4
Clean Down Walls (one side)				
Brick & tile	100 sq ft	2.2	1.1	3.3
Block	100 sq ft	2.0	1.0	3.0
Glased brick & tile	100 sq ft	—	1.5	1.5

Point and caulk manhours are average for this type of work and include the mixing or filling of caulking guns and the placement of same.

Cleaning down wall manhours represents the cleaning down of one side of a wall only using muratic acid or a similar solution.

Manhours do not include scaffolding. See respective table for this charge.

CONVERSION FACTORS FOR VARIOUS BONDS & JOINTS

PERCENTAGE FACTORS TO BE APPLIED

Item	Percent
Type of Bond	
Dutch-Full Header every other course	.18
Dutch-Full Header every sixth course	.25
English-Full Header every other course	.20
English-Full Header every sixth course	.27
Flemish-Full Header every other course	.02
Flemish-Full Header every sixth course	.25
Double Flemish-Full Header every course	.02
Double Flemish-Full Header every sixth course	.15
3-Stretcher Flemish-Full Header every other course	.08
3-Stretcher Flemish-Full Header every sixth course	.10
4-Stretcher Flemish-Full Header every other course	.05
4-Stretcher Flemish-Full Header every sixth course	.10
Type of Joint	
Struck	.02
Weathered	.03
V-Tooled	.05
Concave	.06
Convex	.07
Raked-Out	.05
Rodded	.07
Stripped	.10

Percentages should be added to mason manhours only on masonry tables where the above type of bond or joint is to be used instead of common bond with full headers and flush cut mortar joints.

Section 11

STRUCTURAL STEEL & MISCELLANEOUS IRON

This section includes manhour tables for the erection of structural steel members and the fabrication and erection of various types of miscellaneous structural steel and iron as well as percentage factors to apply against non-ferrous and other metals which may be used for like installations.

The following manhour and percentage tables are based on averages of many projects installed under varied conditions where strict methods and pre-planning were followed and strict reporting of actual costs were recorded in accordance with the notes as appear on the individual table pages.

The listed manhours include time allowance to complete all necessary labor for the particular operation as may be outlined in the various tables.

ERECT STRUCTURAL STEEL – 1 TO 500 TONS

MANHOURS PER TON

Item	Manhours			
	Iron Worker	Oper. Engr.	Oiler	Total
To 20 Tons				
Unload	3.00	1.25	1.25	5.50
Erect, plumb & temporarily bolt	4.50	2.50	2.50	9.50
Fastening:				
Riveting	6.25	1.75	–	8.00
Welding	6.00	–	–	6.00
Bolting by hand	5.00	–	–	5.00
20 to 100 Tons				
Unload	2.00	1.00	1.00	4.00
Erect, plumb & temporarily bolt	4.00	2.25	2.25	8.50
Fastening:				
Riveting	6.00	1.50	–	7.50
Welding	4.50	–	–	4.50
Bolting by hand	4.00	–	–	4.00
100 to 500 Tons				
Unload	1.70	.65	.65	3.00
Erect, plumb & temporarily bolt	3.90	2.05	2.05	8.00
Fastening:				
Riveting	5.25	1.25	–	6.50
Welding	4.00	–	–	4.00
Bolting by hand	3.50	–	–	3.50

Manhours include handling, unloading, shaking out, erecting and fastening of steel as outlined above.

Manhours do not include painting or scaffolding. See respective tables for these charges.

For weights above 500 tons, see manhour table on following page.

ERECT STRUCTURAL STEEL – OVER 500 TONS

MANHOURS PER TON

Item	Manhours			
	Iron Worker	Oper. Engr.	Oiler	Total
500 to 1000 Tons				
Unload	1.70	.65	.65	3.00
Erect, plumb & temporarily bolt	3.40	1.80	1.80	7.00
Fastening:				
Riveting	5.00	1.00	—	6.00
Welding	3.50	—	—	3.50
Bolting by hand	2.50	—	—	2.50
1000 to 5000 Tons				
Unload	1.30	.60	.60	2.50
Erect, plumb & temporarily bolt	3.00	1.75	1.75	6.50
Fastening:				
Riveting	4.40	.60	—	5.00
Welding	2.50	—	—	2.50
Bolting by hand	1.50	—	—	1.50
Over 5000 Tons				
Unload	1.20	.40	.40	2.00
Erect, plumb & temporarily bolt	2.60	1.40	1.40	5.50
Fastening:				
Riveting	4.00	.50	—	4.50
Welding	2.00	—	—	2.00
Bolting by hand	1.00	—	—	1.00

Manhours include handling, unloading, shaking out, erecting and fastening of steel as outlined above.

Manhours do not include painting or scaffolding. See respective tables for these charges.

For weights 500 tons and below, see manhour table on preceding page.

CONCRETE FIREPROOFING FOR COLUMNS AND BEAMS

MANHOURS PER UNITS LISTED

Item	Unit	Manhours					
		Carpenter	Laborer	Iron Worker	Operator Engineer	Oiler	Total
Forms							
Columns	SF	0.14	0.09	–	–	–	0.23
Beams	SF	0.17	0.12	–	–	–	0.29
Mesh Reinforcing							
Sizes 2x2-14/14 thru							
6x6-4/4	SF	–	–	0.03	–	–	0.03
Concrete							
Columns	CY	–	1.86	–	0.15	0.15	2.16
Beams	CY	–	2.15	–	0.17	0.17	2.49

Form manhours include fabrication, erection, shoring, bracing, stripping, and cleaning.

Mesh reinforcing manhours include cutting of mesh and wrapping and securing mesh around columns or beams. It is the average time required for this operation for all sizes as listed.

Concrete manhours include placing with crane and bucket and vibrating of concrete. See Finishing Table under Concrete Section for finishing time requirements.

Hourly fire rating requirements will determine thickness of concrete to be placed.

FIREPROOFING STRUCTURAL STEEL

MANHOURS PER SQUARE YARD

Item Number	Fire Resistance Hours	Manhours		
		Plasterer	Laborer	Total
1	1	0.27	0.14	0.41
2	2	0.48	0.23	0.71
3	4	0.56	0.27	0.83
4	4	0.45	0.22	0.67
5	2	0.37	0.18	0.55
6	2	0.25	0.12	0.37
7	2	0.25	0.12	0.37

The manhours include all operations required for fireproofing structural steel for the listed items as defined below.

Item No. 1: ⅞-inch Portland cement plaster over metal lath tied to ¾-inch vertical channels with 18 gauge tie wire.

Item No. 2: Two separate layers as described in Item No. 1 with air space between the two layers.

Item No. 3: Inner layer of 1-inch vermiculite concrete over paper backed wire fabric wrapped directly around column with an outer layer of 2"x2"–16/16 wire over ¾-inch furring channel and 1-inch vermiculite concrete surface.

Item No. 4: 1½-inch vermiculite gypsum plaster over metal lath wrapped around column and furred 1¼-incn from column flanges. Fill void between lath and flange with plaster.

Item No. 5: 1-inch vermiculite gypsum plaster over self-furring metal lath wrapped directly around column with ⅜-inch space between flange and plaster.

Item No. 6: Four multiple layers of ½-inch gypsum wallboard adhesively secured to column flanges and successive layers. Wallboard layer below outer layer secured to column with doubled 18 gauge ties spaced 15-inches on center.

Item No. 7: Two layers of ⅝-inch type "X" gypsum wallboard screwed to 1⅝"x1"x25 gauge channels spaced 16" on centers and secured to 1½-inch channel furring. Installed parallel to and on each side of beam flanges.

MISCELLANEOUS STRUCTURAL STEEL

MANHOURS PER UNITS LISTED

Item	Unit	Iron Worker	Oper. Engr.	Oiler	Total
Stran Steel Framing					
Stud system	100 sq ft	1.50	–	–	1.50
Rafter system	100 sq ft	3.60	–	–	3.60
Joist system	100 sq ft	2.00	–	–	2.00
Steel Overhead Pipe, Etc., Bridges					
Unload	ton	1.20	.65	.65	2.50
Erect, plumb & temporarily bolt	ton	6.00	3.00	3.00	12.00
Fastening:					
Riveting	ton	3.00	1.00	–	4.00
Welding	ton	2.50	–	–	2.50
Bolting by Hand	ton	2.00	–	–	2.00

Manhours include handling, unloading, shaking out or placing and erecting as outlined above.

Manhours do not include painting or scaffolding. See respective tables for these charges.

DOCK & CANOPY FRAMING, FLOORS & PLATFORMS

MANHOURS PER UNITS LISTED

Item	Unit	Manhours			
		Iron Worker	Oper. Engr.	Oiler	Total
Dock & Canopy Framing					
Unload	ton	1.20	.40	.40	2.00
Erect, plumb & temporarily bolt	ton	4.50	1.50	1.50	7.50
Fastening:					
Riveting	ton	8.25	2.00	—	10.25
Welding	ton	2.10	—	—	2.10
Bolting by Hand	ton	1.50	—	—	1.50
Floors & Platforms					
Unload	ton	2.00	.60	.60	3.20
Erect & Fasten:					
Platform Framing	ton	31.00	9.00	—	40.00
Catwalk Framing	ton	30.00	8.00	—	38.00
Checkered floor plate	cwt	2.00	—	—	2.00

Manhours include handling, unloading, shaking out or placing and erecting of steel for the items as outlined above.

Manhours do not include painting or scaffolding. See respective tables for these charges.

Operating Engineer Manhours for floors and platforms is that of hoist engineer.

BAR JOIST, PARTITION FRAMING, MONORAIL & EQUIPMENT SUPPORTS

MANHOURS PER UNITS LISTED

Item	Unit	Manhours			
		Iron Worker	Oper. Engr.	Oiler	Total
Bar Joists					
Unload	ton	1.50	.75	.75	3.00
Erect & fasten	ton	8.00	2.00	2.00	12.00
Partition Framing					
Unload	cwt	.15	.05	.05	.25
Fabricate	cwt	2.00	—	—	2.00
Erect & fasten	cwt	1.25	—	—	1.25
Monorails					
Unload	cwt	.15	.05	.05	.25
Erect & fasten	cwt	1.25	—	—	1.25
Equipment Supports					
Unload	cwt	.15	.05	.05	.25
Fabricate	cwt	2.50	—	—	2.50
Erect & fasten	cwt	2.75	—	—	2.75

Manhours include unloading, handling, hauling, fabricating, placing and erecting of Items as outlined above.

Manhours do not include painting or scaffolding. See respective tables for these charges.

If crafts other than those listed above should install any of the above items due to craft jurisdiction, simply substitute that craft for those listed.

PIPE AND SQUARE TUBE COLUMNS

MANHOURS PER 12' LENGTH

Item	Manhours			
	Iron worker	Operator Engineer	Oiler	Total
Pipe Columns With Cap and Base				
Erect— 3" round	0.75	0.20	0.20	1.15
Erect—3½" round	0.75	0.25	0.25	1.25
Erect— 4" round	0.80	0.25	0.25	1.30
Erect— 5" round	0.90	0.25	0.25	1.40
Erect— 6" round	0.95	0.25	0.25	1.45
Erect— 8" round	1.00	0.25	0.25	1.50
Erect—10" round	1.05	0.25	0.25	1.55
Erect—12" round	1.10	0.25	0.25	1.60
Square Tube Columns with Cap and Base				
Erect—4"x4" square	0.90	0.25	0.25	1.40
Erect—5"x5" square	0.95	0.25	0.25	1.45
Erect—6"x6" square	1.00	0.25	0.25	1.50
Erect—7"x7" square	1.05	0.25	0.25	1.55
Erect—8"x8" square	1.10	0.25	0.25	1.60

Manhours include unloading, handling, shaking out, and erecting of up to 12'0" long columns including attached cap and base plate.

Pipe Column manhours are for any schedule pipe through double extra strong.

Square Tube Column manhours are for plate wall thickness of ¼" through ½".

For lengths greater than 12'0", increase manhours ten (10) percent for each additional five feet or portion thereof.

Manhours do not include setting anchor bolts, grouting, painting, or scaffolding. See respective tables for these charges.

STEEL GRATING

MANHOURS PER UNITS LISTED

Item	Weight lbs/sq ft	Unit	Iron Worker Manhours
Grating Bar Size			
¼"x3/16"	5.9	100 SF	11.00
1"x¼"	5.2	100 SF	10.00
1"x3/16"	7.6	100 SF	14.00
1¼"x¼"	6.3	100 SF	12.00
1¼"x3/16"	9.2	100 SF	17.00
1½"x¼"	7.4	100 SF	14.00
1½"x3/16"	10.9	100 SF	18.00
1¾"x3/16"	12.6	100 SF	19.00
2"x3/16"	14.3	100 SF	20.00
2¼"x3/16"	15.9	100 SF	22.00
2½"x3/16"	17.6	100 SF	24.00
Straight and Diagonal Cutting	–	LF	0.10
Circular Cutting	–	LF	0.15
Straight and Diagonal Banding	–	LF	0.12
Circular Banding	–	LF	0.20
Checker Plate or Angle Nosing Welded to Grating	–	LF	0.25
Abrasive Nozing Attached to Grating	–	LF	0.35

Grating weights are based on bearing bars at 1-3/16" on center and cross bars at 4" on center.

Manhours include fabrication where required and installation of items as listed.

Decrease manhours twenty-five (25) percent for the installation of aluminum or plastic grating of the same size.

Manhours do not include printing or scaffolding. See respective tables for these time frames.

UNLOAD MISCELLANEOUS STEEL ITEMS— FABRICATE AND ERECT STAIRS AND LADDERS

MANHOURS PER UNITS LISTED

Item	Unit	Iron Worker	Operator Engineer	Oiler	Total
		Manhours			
Unload Miscellaneous Steel Items	Ton	1.80	0.60	0.60	3.00
Stairs					
Field fabricate—all sizes	CWT	3.50	–	–	3.50
Field erect—3'0" wide	LF	0.35	0.15	0.15	0.65
Field erect—3'6" wide	LF	0.45	0.15	0.15	0.75
Field erect—4'0" wide	LF	0.55	0.15	0.15	0.85
Field fabricate stair platform	CWT	2.00	–	–	2.00
Erect rectangular platforms	SF	0.10	0.05	0.05	0.20
Erect circular platforms	SF	0.15	0.05	0.05	0.25
Ladders					
Fabricate straight ladders	CWT	6.00	–	–	6.00
Fabricate ships ladder	CWT	5.00	–	–	5.00
Fabricate safety cages	CWT	6.00	–	–	6.00
Erect straight ladders	LF	0.20	0.05	0.05	0.30
Erect ships ladder	LF	0.25	–	–	0.25
Erect safety cage	LF	0.20	0.05	0.05	0.30
Erect ¼" ladder rungs	EA	0.50	–	–	0.50

Unloading manhours include handling, unloading, and shaking out of miscellaneous steel items such as outlined above.

Erection manhours include complete field erection of the above items.

Manhours do not include painting or scaffolding. See respective tables for these time requirements.

PIPE AND ANGLE HANDRAILS AND TOE PLATES

MANHOURS PER UNITS LISTED

Item	Unit	Iron Worker Manhours
Standard Pipe Handrail—3'6" High		
Fabricate 1¼" & 1½"—straight runs	CWT	14.70
Fabricate 1¼" & 1½"—angled runs	CWT	17.60
Erect 1¼" pipe—straight free standing	LF	0.16
Erect 1¼" pipe—angled free standing	LF	0.19
Erect 1½" pipe—straight free standing	LF	0.20
Erect 1½" pipe—angled free standing	LF	0.24
Single Pipe Handrail—Wall Attached		
Fabricate 1¼" & 1½" pipe	CWT	6.60
Erect 1¼" pipe	LF	0.12
Erect 1½" pipe	LF	0.16
Erect wall brackets	EA	0.50
Standard Angle Iron Handrail—3'6" High		
Fabricate—straight runs	CWT	35.00
Fabricate—angled runs	CWT	42.00
Erect 2"x2"x¼"—straight free standing	LF	0.20
Erect 2"x2¼"—angled free standing	LF	0.23
Erect 2½"x2½"x¼"—straight free standing	LF	0.24
Erect 2½"2½"x¼"—angled free standing	LF	0.28
Standard Toe or Kick Plate		
10 ga x 6"	LF	0.15
3/16"x6"	LF	0.20
¼"x6"	LF	0.25

Fabrication manhours include shaking out and fabricating a standard two-rail handrail or single plate toe plate.

Straight runs are those with all connecting members at 90° angles. Angled runs are those with all connecting members at angles other than 90°.

Erection manhours include the field handling and complete erection of the prefabricated item.

Manhours do not include the placement of sleeves or painting. See respective tables for these time frames.

For erection of handrails made of light tubes and flat bar, use same manhours as listed for angle handrails. For handrails with expanded metal panels, increase angle iron manhours thirty (30) percent.

MISCELLANEOUS IRON & STEEL

MANHOURS PER HUNDRED (100) POUNDS

Item	Manhours			
	Iron Worker	Oper. Engr.	Oiler	Total
Curb Angles				
Unload	.09	.03	.03	.15
Fabricate	2.40	—	—	2.40
Anchor Bolts				
Fabricate	2.00	—	—	2.00
Steel Curbing				
Unload	.15	.05	.05	.25
Fabricate	2.50	—	—	2.50
Install	1.80	—	—	1.80
Pipe Sleeves				
Fabricate	3.00	—	—	3.00
Loose Lintels - 1/4", 5/16" & 3/8"				
Unload	.15	.05	.05	.25
Fabricate	1.50	—	—	1.50
Install	2.00	.50	—	2.50
Ledger Angles				
Unload	.15	.05	.05	.25
Fabricate	2.00	—	—	2.00
Install	2.50	.50	—	3.00

Manhours include handling, unloading, placing, fabricating, and erecting of items as outlined and described above.

For the installation of curb angles, anchor bolts and pipe sleeves see tables under Section 8 entitled, "Miscellaneous Embedded Items."

Manhours do not include painting or scaffolding. See respective tables for these charges.

DOOR, LOUVER & DUCT FRAMES

MANHOURS PER HUNDRED (100) POUNDS

| Item | Manhours | | | |
	Iron Worker	Oper. Engr.	Oiler	Total
Door Frames				
Unload	.15	.05	.05	.25
Fabricate	2.30	–	–	2.30
Install	2.25	–	–	2.25
Louver Frames				
Unload	.15	.05	.05	.25
Fabricate	2.25	–	–	2.25
Erect	1.60	.40	–	2.00
Duct Frames				
Unload	.15	.05	.05	.25
Fabricate	2.00	–	–	2.00
Erect	2.50	–	–	2.50

Manhours include handling, unloading, fabricating, and placing of miscellaneous steel items as outlined above.

Manhours do not include painting or scaffolding. See respective tables for these charges.

MISCELLANEOUS STEEL ITEMS

MANHOURS PER UNITS LISTED

Item	Unit	Iron Worker	Oper. Engr.	Oiler	Total
Miscellaneous Hangers					
Fabricate	cwt	2.50	–	–	2.50
Erect	cwt	3.00	–	–	3.00
Metal Thresholds					
Install	each	1.50	–	–	1.50
Wheel Guards					
Install	pc	.50	–	–	.50
Bins & Hoppers					
Unload	cwt	.15	.05	.05	.25
Fabricate	cwt	1.50	–	–	1.50
Erect & bolt – no welding	cwt	.50	.05	.05	.60
Trench Framing					
Unload	cwt	.15	.05	.05	.25
Fabricate	cwt	2.25	–	–	2.25
Install	cwt	1.75	–	–	1.75
Trench Covers					
Install Plate	100 sq ft	10.00	–	–	10.00
Install Grating	100 sq ft	14.00	–	–	14.00

Manhours include handling, unloading, placing, fabricating and installing items as outlined above.

Manhours do not include painting or scaffolding. See respective tables for these charges.

COMPARISON PERCENTAGE – OTHER METALS TO MILD STEEL

PERCENTAGES

Metals	Percentage Factors	
	Burn or Weld	Hacksaw, Grind or File
Cast Iron	2.00	1.75
Cast Steel	1.90	1.25
Stainless Steel	1.75	3.40
Nickel	1.40	3.25
Monel	1.75	3.25
Copper	1.15	.50
Cast Brass	1.25	.75
Brass	1.20	.45
Bronze	1.25	.75
Aluminum	.60	.40

The above percentages are for the purpose of application against the preceding structural steel and miscellaneous iron manhour tables for the same type of work using the materials outlined above.

Section 12

CARPENTRY & MILLWORK

This section covers only items of carpentry and millwork which might be encountered in heavy commercial and industrial buildings.

The following manhour tables are averages of several projects and are based on first class workmanship.

Truck Driver hauling time has been allowed throughout the various tables for the hauling of materials from a centrally located storage or fabricating yard to erection site. Should materials be delivered directly from the vendor's storage yard or warehouse direct to the site of erection, the estimator should eliminate these charges from the tables. Should laborers carry lumber a short distance on foot, from storage to erection site, then labor time should be substituted for truck driver time.

STRUCTURAL LUMBER

MANHOURS PER THOUSAND (1000) FOOT BOARD MEASURE

Item	Manhours					
	Carpenter	Laborer	Truck Driver	Oper. Engt.	Oiler	Total
Sills	15.00	3.75	.35	–	–	19.10
Bracing	18.75	4.50	.35	–	–	23.60
Floor Joist	12.75	5.25	.35	–	–	18.35
Roof Joist	13.60	4.00	.35	–	–	17.95
Beams	7.50	5.25	.35	–	–	13.10
Columns & Posts	25.50	6.75	.35	–	–	32.60
Girders	9.38	7.13	.35	–	–	16.86
Ordinary Rafters	17.25	4.50	.35	–	–	22.10
Hip & Valley Rafters	27.20	4.80	.35	–	–	32.25
Bridging & Blocking	35.75	6.50	.35	–	–	42.60
Trusses	30.00	6.40	.50	1.25	1.25	39.40
Purlins	13.60	4.00	.35	–	–	17.95
Nailers	24.00	7.70	.35	–	–	32.05

Manhours include handling at storage or saw yard, hauling up to 1000 feet to erection site, fabrication and erection of items as outlined above.

Manhours do not include installation of studs, plates, sheathing or scaffolding. See respective tables for these charges.

METAL BASES AND CAPS FOR WOOD COLUMNS, POSTS, AND BEAMS

MANHOURS PER HUNDRED (100)

Item	Sizing	Manhours			
		Carpenter	Laborer	Truck Driver	Total
Column or Post Base Plates	**Column or Post Size**				
"U" type strap	4"x4" through 6"x6"	17.5	4.5	2.0	24.0
"U" type strap	6"x8" through 10"x10"	22.0	6.0	2.0	30.0
Adjustable type	4"x4" through 6"x6"	14.0	4.0	2.0	20.0
Elevated type	4"x4" through 4"x6"	29.0	8.0	3.0	40.0
Elevated type	6"x6"	35.00	10.0	3.0	48.0
Caps and Straps	**Post Size x Beam Size**				
Columns	3⅝"x3⅝"	29.0	8.0	3.0	40.0
Columns	3⅝"x3¼"	29.0	8.0	3.0	40.0
Columns	3⅝"x5½"	35.0	10.0	3.0	48.0
Columns	5½"x3⅝"	35.0	10.0	3.0	48.0
Columns	5½"x5½"	41.5	11.5	3.0	56.0
Columns	5½"x5¼"	41.5	11.5	3.0	56.0
Columns	7½"x5¼"	42.0	12.0	4.0	58.0
Columns	7½"x7½"	42.0	12.0	4.0	58.0
End or intermediate post	All sizes	22.0	6.0	2.0	30.0
Three-way post	All sizes	19.0	5.0	2.0	26.0
Four-way post	All sizes	28.0	8.0	2.0	38.0
"T" and "L" straps	All sizes	10.0	3.0	1.0	14.0

Column or post base plates manhours include time required for laying out and placing base prior to pouring concrete.

"U" type strap and elevated base plate manhours include drilling post and bolting base to post.

Adjustable base plate manhours include nailing base plate to post.

Column cap and "T" and "L" strap manhours include drilling holes in columns and beams and bolting columns and beams to cap or strap.

End or intermediate, three-way, and four-way post cap manhours include nailing caps to post and beam.

All manhours include job procurement and hauling up to 1000 feet of erection site.

Manhours do not include installation of column, post, or beam. See respective table for these time frames.

METAL JOIST HANGERS

MANHOURS PER HUNDRED (100)

Joist Size	Manhours			
	Carpenter	Laborer	Truck Driver	Total
Prong Joist Hangers				
1⅛"x4" & 6"	3.5	1.5	0.6	5.6
1⅛"x8" & 10"	6.5	2.8	1.1	10.4
1⅛"x12". 14" & 16"	9.6	4.1	1.7	15.4
Heavy Duty Joist Hangers				
2"x4"	2.6	1.1	0.5	4.2
2"x6"	3.0	1.3	0.5	4.8
2"x8"	4.1	1.7	0.7	6.5
2"x10"	4.9	2.1	0.8	7.8
2"x12"	6.0	2.6	1.0	9.6
2"x14"	6.8	2.9	1.1	10.8
3"x4"	2.6	1.1	0.5	4.2
3"x6"	4.9	2.1	0.8	7.8
3"x8"	5.7	2.4	1.0	9.1
3"x10"	7.5	3.2	1.3	12.0
3"x12"	8.3	3.5	1.4	13.2
3"x14"	9.8	4.2	1.6	15.6
3"x16"	10.5	4.5	1.8	16.8
4"x4"	2.6	1.1	0.5	4.2
4"x6"	4.9	2.1	0.8	7.8
4"x8"	5.7	2.4	1.0	9.1
4"x10"	7.5	3.2	1.3	12.0
4"x12"	8.3	3.5	1.4	13.2
4"x14"	9.8	4.2	1.6	15.6
4"x16"	10.5	4.5	1.8	16.8
6"x6"	4.9	2.1	0.8	7.8

Manhours include job procurement. hauling up to 100 feet. and installing and nailing joist hanger in position.

Manhours do not include installation of joist. See respective table for this time frame.

WOOD FRAMING

MANHOURS PER UNITS LISTED

Item	Unit	Manhours			
		Carpenter	Laborer	Truck Driver	Total
Exterior					
Plates	Mfbm*	15.00	3.75	.35	19.10
Studs	Mfbm*	16.50	4.50	.35	21.35
Sheathing laid horizontally	Mfbm*	12.75	2.25	.35	15.35
Sheathing laid diagonally	Mfbm*	15.00	3.00	.35	18.35
Sheathing - plywood	100 sq ft	1.00	.25	.05	1.30
Interior					
Plates	Mfbm*	15.00	3.75	.35	19.10
Studs	Mfbm*	18.75	4.50	.35	23.60
Sheathing laid horizontally	Mfbm*	12.75	2.25	.35	15.35
Sheathing laid diagonally	Mfbm*	15.00	3.00	.35	18.35
Sheathing - plywood	100 sq ft	1.00	.25	.05	1.30

Manhours include handling, hauling up to 1000 feet, fabricating of studs and plates with power saws, complete installation and removal of excess sheathing with power saws.

Manhours do not include placement of beams, joists, rafters, roof deck or scaffolding. See respective tables for these charges.

*1,000 foot board measure

LOADING DOCKS & SUB FLOORING

MANHOURS PER UNITS LISTED

Item	Unit	Manhours			
		Carpenter	Laborer	Truck Driver	Total
Loading Docks					
Framing	Mfbm*	15.00	3.00	.35	18.35
Roof decking - lumber	Mfbm*	10.50	4.50	.35	15.35
Roof decking - plywood	100 sq ft	1.00	.50	.05	1.55
Bumpers	Mfbm*	36.00	18.00	.35	54.35
Sub-Flooring					
T & G laid regular	Mfbm*	12.00	3.38	.35	15.73
T & G laid diagonal	Mfbm*	13.50	3.38	.35	17.23
Plywood sub-flooring	100 sq ft	.75	.25	.05	1.05
Plank Flooring	Mfbm*	8.63	3.00	.35	11.98

Manhours include necessary handling, hauling up to 1000 feet, fabricating and erecting of items as outlined above.

Manhours do not include painting or carpentry operations other than those listed above.

*1,000 foot board measure

GLUED LAMINATED BEAMS-HANGERS AND CONNECTORS

MANHOURS REQUIRED EACH

Beam Size Width X Depth	Manhours														
	1			2			3			4			5		
	C	L	T	C	L	T	C	L	T	C	L	T	C	L	T
3⅛"x13½"	–	–	–	0.13	0.07	0.20	0.15	0.08	0.23	–	–	–	0.08	0.04	0.12
3⅛"x15"	0.05	0.03	0.08	0.13	0.07	0.20	0.15	0.08	0.23	–	–	–	0.08	0.04	0.12
3⅛"x16½"	0.05	0.03	0.08	0.13	0.07	0.20	0.15	0.08	0.23	–	–	–	0.08	0.04	0.12
5⅛"x15"	0.07	0.03	0.10	0.13	0.07	0.20	0.17	0.09	0.26	0.38	0.18	0.56	0.08	0.04	0.12
5⅛"x16½"	0.07	0.03	0.10	0.13	0.07	0.20	0.17	0.09	0.26	0.38	0.18	0.56	0.08	0.04	0.12
5⅛"x18"	0.09	0.04	0.13	0.15	0.07	0.22	0.17	0.09	0.26	0.44	0.22	0.66	0.09	0.05	0.14
5⅛"x19½"	0.09	0.04	0.13	0.15	0.07	0.22	0.20	0.10	0.30	0.44	0.22	0.66	0.09	0.05	0.14
5⅛"x21"	0.11	0.05	0.16	0.15	0.07	0.22	0.20	0.10	0.30	0.44	0.22	0.66	0.09	0.05	0.14
5⅛"x22½"	0.11	0.05	0.16	0.15	0.07	0.22	0.20	0.10	0.30	0.48	0.24	0.72	0.09	0.05	0.14
5⅛"x24"	0.12	0.06	0.18	0.16	0.08	0.24	0.21	0.11	0.32	0.48	0.24	0.72	0.11	0.05	0.16
5⅛"x25½"	0.12	0.06	0.18	0.16	0.08	0.24	0.21	0.11	0.32	0.48	0.24	0.72	0.11	0.05	0.16
6¾"x24"	0.12	0.06	0.18	0.16	0.08	0.24	0.21	0.11	0.32	0.62	0.30	0.92	–	–	–
6¾"x25½"	0.12	0.06	0.18	0.16	0.08	0.24	0.21	0.11	0.32	0.62	0.30	0.92	–	–	–
6¾"x27"	0.13	0.07	0.20	0.17	0.08	0.25	0.21	0.11	0.32	0.62	0.30	0.92	–	–	–
6¾"x28½"	0.13	0.07	0.20	0.17	0.08	0.25	0.23	0.12	0.35	0.67	0.33	1.00	–	–	–
6¾"x30"	0.13	0.07	0.20	0.18	0.09	0.27	0.23	0.12	0.35	0.67	0.33	1.00	–	–	–
6¾"x31½"	–	–	–	0.18	0.09	0.27	0.23	0.12	0.35	0.67	0.33	1.00	–	–	–
8¾"x31½"	–	–	–	0.18	0.09	0.27	0.34	0.16	0.50	0.86	0.42	1.28	–	–	–
8¾"x33"	–	–	–	0.18	0.09	0.27	0.34	0.16	0.50	0.86	0.42	1.28	–	–	–
8¾"x34½"	–	–	–	0.18	0.09	0.27	0.34	0.16	0.50	0.91	0.45	1.36	–	–	–
8¾"x36"	–	–	–	0.20	0.10	0.30	0.44	0.20	0.64	0.91	0.45	1.36	–	–	–
8¾"x37½"	–	–	–	0.20	0.10	0.30	0.44	0.20	0.64	0.91	0.45	1.36	–	–	–

Codes

1 = joist and perlin hangers

2 = ledger hangers

3 = saddle hangers

4 = hinge connectors (to greater dimensioned beam ends.)

5 = cantilever hinge connector

C = carpenter manhours

L = laborer manhours

T = total mahours

Manhours are for handling and installing of hanger or connector to the support structure only.

Beam seats for concrete or masonry pilasters can be installed for approximately the same manhours as required for ledger hangers.

Manhours do not include beam erection or fastening. See respective table for this time frame.

INSTALL LAMINATED BEAMS WITH STRUCTURAL FASTENERS

MANHOURS REQUIRED EACH

Beam Size Width X Depth (in.)	Beam Length (ft.)	Beam Weight (lbs.)	Manhours				
			Carpenter	Laborer	Operator Engineer	Power Tool Operator	Total
3⅛x9	16	106	0.13	0.07	0.03	0.07	0.30
3⅛x10½	16	123	0.13	0.07	0.03	0.07	0.30
3⅛x12	16	142	0.13	0.07	0.03	0.07	0.30
3⅛x13½	16	160	0.15	0.08	0.03	0.08	0.34
3⅛x15	16	178	0.16	0.08	0.03	0.08	0.35
3⅛x16½	16	195	0.18	0.09	0.04	0.09	0.40
3⅛x15	24	266	0.20	0.10	0.04	0.10	0.45
3⅛x16½	24	293	0.22	0.11	0.06	0.11	0.50
5⅛x15	24	434	0.24	0.12	0.07	0.12	0.55
5⅛x16½	24	480	0.24	0.12	0.07	0.12	0.55
5⅛x18	24	523	0.26	0.13	0.08	0.13	0.60
5⅛x19½	24	566	0.26	0.13	0.08	0.13	0.60
5⅛x21	24	610	0.30	0.15	0.08	0.15	0.68
5⅛x16½	32	640	0.30	0.15	0.08	0.15	0.68
5⅛x18	32	698	0.32	0.16	0.08	0.16	0.72
5⅛x19½	32	755	0.33	0.16	0.10	0.16	0.75
5⅛x21	32	813	0.33	0.17	0.10	0.17	0.77
5⅛x22½	32	870	0.33	0.17	0.10	0.17	0.77
5⅛x24	32	928	0.35	0.17	0.11	0.17	0.80
5⅛x25½	32	963	0.35	0.17	0.11	0.17	0.80
6¾x24	50	1910	0.40	0.19	0.12	0.19	0.90
6¾x25½	50	2030	0.40	0.20	0.12	0.20	0.92
6¾x27	50	2150	0.41	0.20	0.13	0.20	0.94
6¾x28½	50	2270	0.43	0.21	0.13	0.21	0.98
6¾x30	50	2390	0.44	0.21	0.14	0.21	1.00
6¾x31½	50	2510	0.46	0.22	0.15	0.22	1.05
8¾x31½	50	3250	0.48	0.23	0.16	0.23	1.10
8¾x33	50	3405	0.53	0.26	0.16	0.26	1.21
8¾x34½	50	3560	0.55	0.27	0.16	0.27	1.25
8¾x36	50	3715	0.57	0.28	0.17	0.28	1.30
8¾x37½	50	3870	0.59	0.29	0.18	0.29	1.35

Manhours include handling, unloading at erection site, rigging, picking, setting, aligning, and fastening in place with structural fasteners.

Manhours do not include installation of hangers or connectors to structural supports. See respective table for this time frame.

INSTALL LAMINATED BEAMS WITH TWO MACHINE BOLTS EACH END

MANHOURS REQUIRED EACH

Beam Size Width X Depth (in.)	Beam Length (ft.)	Beam Weight (lbs.)	Manhours				
			Carpenter	Laborer	Operator Engineer	Power Tool Operator	Total
3⅛x9	16	106	0.16	0.08	0.04	0.62	0.90
3⅛x10½	16	123	0.16	0.08	0.04	0.62	0.90
3⅛x12	16	142	0.16	0.08	0.04	0.62	0.90
3⅛x13½	16	160	0.18	0.10	0.04	0.70	1.02
3⅛x15	16	178	0.19	0.10	0.04	0.72	1.05
3⅛x16½	16	195	0.22	0.11	0.05	0.82	1.20
3⅛x15½	24	266	0.24	0.12	0.05	0.94	1.35
3⅛x16½	24	293	0.26	0.13	0.07	0.94	1.40
5⅛x15	24	434	0.29	0.14	0.08	0.94	1.45
5⅛x16½	24	480	0.29	0.14	0.08	0.94	1.45
5⅛x18	24	523	0.30	0.15	0.10	0.95	1.50
5⅛x19½	24	566	0.30	0.15	0.10	0.95	1.50
5⅛x21	24	6.10	0.35	0.17	0.10	0.97	1.59
5⅛x16½	32	640	0.35	0.17	0.10	0.97	1.59
5⅛x18	32	698	0.38	0.19	0.10	1.01	1.68
5⅛x19½	32	755	0.40	0.19	0.12	1.05	1.76
5⅛x21	32	813	0.40	0.20	0.12	1.08	1.80
5⅛x22½	32	870	0.40	0.20	0.12	1.08	1.80
5⅛x24	32	928	0.42	0.20	0.13	1.12	1.87
5⅛x25½	32	963	0.42	0.20	0.13	1.12	1.87
6¾x24	50	1910	0.48	0.23	0.14	1.26	2.11
6¾x25½	50	2030	0.48	0.24	0.14	1.29	2.15
6¾x27	50	2150	0.49	0.24	0.16	1.31	2.20
6¾x28½	50	2270	0.52	0.25	0.16	1.32	2.25
6¾x30	50	2390	0.53	0.25	0.17	1.35	2.30
6¾x31½	50	2510	0.55	0.26	0.18	1.35	2.34
8¾x31½	50	3250	0.56	0.26	0.18	1.38	2.38
8¾x33	50	3405	0.56	0.26	0.18	1.42	2.42
8¾x34½	50	3560	0.60	0.30	0.18	1.42	2.50
8¾x36	50	3715	0.64	0.32	0.20	1.44	2.60
8¾x37½	50	3870	0.68	0.33	0.22	1.47	2.70

Manhours include handling, unloading at erection site, rigging, picking, setting, aligning, boring two holes at each end of beam for machine bolts and fastening into position with machine bolts.

Manhours do not include installation of hangers or connectors to structural supports. See respective table for this time frame.

ROOF LUMBER

MANHOURS PER UNITS LISTED

Item	Unit	Manhours			
		Carpenter	Laborer	Truck Driver	Total
T & G Lumber Roof Decking	Mfbm*	12.60	5.40	.34	18.34
Plywood Roof Decking	100 sq ft	.90	.16	.05	1.11
Stiffeners	Mfbm*	44.20	7.20	.40	51.80
Edging	Mfbm*	48.00	8.00	.40	56.40
Saddles	Mfbm*	16.00	5.60	.35	21.95
Monotor Eaves	Mfbm*	45.00	7.50	.40	52.90
Cant Strip	100 lin ft	1.58	—	—	1.58
Wood Curbs	Mfbm*	15.00	5.25	.35	20.60
Skylight Framing	Mfbm*	28.80	6.40	.35	35.55
Vent Framing	Mfbm*	33.00	7.50	.35	40.85
Monotor Framing	Mfbm*	21.60	7.20	.35	29.15
Wood Coping	100 lin ft	4.95	1.80	—	7.75
Wood Facia	Mfbm*	45.00	7.50	.35	52.85
Soffitts	Mfbm*	45.50	7.75	.35	53.60
Catwalks or Roof Walkways	Mfbm*	9.00	8.10	.35	17.45

Manhours include handling at storage or saw yard, hauling up to 1000 feet to erection site, fabrication and erection.

Manhours do not include framing, roofing or scaffolding. See respective tables for these charges.

*1,000 foot board measure

WALL COVERINGS

MANHOURS PER UNITS LISTED

Item	Unit	Manhours			
		Carpenter	Laborer	Truck Driver	Total
Drop Siding - Lap & Bevel	Mfbm*	19.13	3.38	.35	22.86
Building Paper	100 sq ft	1.00	—	—	1.00
Siding Shingles	100 sq ft	3.50	1.25	.12	4.87
Fiberboard	100 sq ft	3.00	1.00	.15	4.15
Gypsum Board	100 sq ft	2.40	1.00	.15	3.55
Asbestos Cement Board	100 sq ft	4.50	1.00	.15	5.65
Plaster Board	100 sq ft	2.40	.75	.15	3.30
Sheet Rock	100 sq ft	1.50	.50	.15	2.15
Tape Joint System	100 lin ft	3.50	—	.15	3.65
Plywood Paneling	100 sq ft	4.50	1.00	.15	5.65

Manhours include handling, hauling up to 1000 feet and erecting of items outlined above.

See other manhour tables for items not listed above.

*1,000 foot board measure

LAYING & FINISHING FLOORS

MANHOURS PER UNITS LISTED

Item	Unit	Manhours				
		Carpenter	Laborer	Floor Sander	Truck Driver	Total
Laying Flooring						
Maple	mfbm*	18.40	4.00	—	.35	22.75
Oak	mfbm*	20.40	4.80	—	.35	25.55
Birch	mfbm*	20.40	4.80	—	.35	25.55
Pine	mfbm*	14.40	4.00	—	.35	18.75
Fir	mfbm*	16.80	4.00	—	.35	21.15
Finish Hardwood Floors						
Sanding	100 sq ft	—	—	1.80	—	1.80
Finishing	100 sq ft	—	—	3.00	—	3.00
Finish Softwood Floors						
Sanding	100 sq ft	—	—	1.20	—	1.20
Finishing	100 sq ft	—	—	2.50	—	2.50
Refinish Old Floors						
Sanding	100 sq ft	—	—	3.50	—	3.50
Finishing	100 sq ft	—	—	3.40	—	3.40

Laying floor units include necessary manhours as may be required for the handling, job hauling and placement of the various types of finished flooring as outlined above.

Finish flooring units include manhours for the complete sanding, filling and glazed or waxed finishing.

Manhours do not include the placement of sub-flooring.   See respective table for this charge.

*1,000 foot board measure

FURRING, GROUNDS, BLOCKING & CAULKING

MANHOURS PER HUNDRED (100) LINEAR FEET

Item	Manhours			
	Carpenter	Laborer	Truck Driver	Total
Wood Furring				
Wall furring strips	2.80	–	.01	2.81
Floor furring strips	1.50	–	.01	1.51
Grounds & Blocking				
On brick	4.00	.50	.01	4.51
For plaster	4.00	.50	.01	4.51
Caulking Openings				
With pressure gun	1.60	–	–	1.60
Without pressure gun	2.70	–	–	2.70

Manhours cover handling, hauling up to 1000 feet, fabricating and erecting, including installation of insert plugs, as may be necessary for wood furring, grounds and blocking.

Caulking manhours include loading and caulking with gun or hand and putty knife placing.

Manhours do not include scaffolding. See respective table for this charge.

EXTERIOR TRIM

MANHOURS PER UNITS LISTED

Item	Unit	Manhours			
		Carpenter	Laborer	Truck Driver	Total
Cornice - Two Member	100 lin ft	4.80	1.00	.03	5.83
Cornice - Three Member	100 lin ft	6.80	1.00	.03	7.83
Corner Board	100 lin ft	3.20	.25	.04	3.49
Verge Board	100 lin ft	10.00	.15	.04	10.19
Miscellaneous Mouldings	100 lin ft	4.00	1.00	.02	5.02
Blinds or Shutters	pair	1.25	.40	.05	1.70
Wood Threshold	each	.50	—	.03	.53
Porch Columns (average)	each	1.50	.25	.10	1.85
Porch Rail & Baluster	100 lin ft	33.75	4.00	.10	37.85
Matched & Beaded Ceiling	100 sq ft	3.00	.50	.15	3.65
Steps - Average 4-High	set	9.75	1.00	.15	10.90

Manhours include handling, hauling from storage up to 1000 feet, fabrication and erection of items as listed above for exterior construction.

Manhours do not include scaffolding. See respective table for this charge.

INTERIOR TRIM

MANHOURS PER UNITS LISTED

Item	Unit	Manhours			
		Carpenter	Laborer	Truck Driver	Total
Baseboard - Two Member	100 lin ft	4.80	1.00	.10	5.90
Chair Rail	100 lin ft	3.20	1.00	.05	4.25
Picture Molding	100 lin ft	2.80	.75	.05	3.60
Beam Casing	100 lin ft	4.00	1.00	.10	5.10
Miscellaneous Modlings	100 lin ft	2.80	.75	.05	3.60
Assembled Wood Wainscot	100 sq ft	3.20	1.00	.15	4.35
Knocked Down Wood Paneling	100 sq ft	7.20	1.50	.15	8.85
Door Saddles	each	.20	—	.02	.22
Door Casings	100 lin ft	4.00	.25	.05	4.30
Door Stops	opening	.75	—	.03	.78
Window Casings	100 lin ft	4.00	.25	.05	4.30
Window Stools & Apron	100 lin ft	1.00	.25	.05	1.55
Window Stops	100 lin ft	3.35	—	.03	3.38
Wood Handrail	100 lin ft	16.00	1.00	.15	17.15
Prefabbed Wood Stairs (avg.)	flight	9.00	1.00	.25	10.25

Manhours include handling, hauling from storage up to 1000 feet, fabrication and erection of items as listed above for interior construction.

Manhours do not include scaffolding. See respective table for this charge.

WALL, CEILING & FLOOR INSULATION

MANHOURS PER HUNDRED (100) SQUARE FEET

Item		Manhours		
		Carpenter	Truck Driver	Total
Flexible Roll		1.20	.05	1.25
Foam Glass		1.60	.05	1.65
Vapor Seal Paper		.25	.05	.30
Insulating Lath		.90	.05	.95
Rockwool Batts		1.25	.05	1.30
Strip Rockwool		1.00	.05	1.05
Loose Rockwool		1.90	.05	1.95
Granular Mineral		1.10	.05	1.15

Manhours include handling, hauling and placing of insulation, outlined above, for walls, ceilings and floors.

Manhours do not include roof insulation or scaffolding. See respective tables for these charges.

EXTERIOR WOOD DOORS & TRANSOMS

MANHOURS PER UNITS LISTED

Item	Unit	Manhours		
		Carpenter	Truck Driver	Total
Single Swing Doors				
Rough Buck	opening	1.80	.05	1.85
Frame	opening	1.40	.05	1.45
Hang door	each	1.00	.05	1.05
Trim	opening	1.00	.03	1.03
Double Swing Doors				
Rough buck	opening	2.70	.06	2.76
Frame	opening	1.60	.06	1.66
Hang door	pair	1.80	.06	1.86
Trim	opening	1.70	.05	1.75
Screen Doors & Hardware	each	2.20	.10	2.30
Sliding Doors & Hardware	sq ft	.30	.005	.305
Overhead Doors & Hardware	sq ft	.25	.005	.255
Transoms				
Rough buck	opening	.60	.04	.64
Frame	opening	.25	.03	.28
Hang sash	opening	1.00	.03	1.03
Trim	opening	.60	.03	.63

Manhours include handling, hauling from job storage and installing exterior doors and transoms, as outlined above.

Manhours do not include the placement of hardware or glass and glazing. See respective tables for these charges.

INTERIOR WOOD DOORS & TRANSOMS

MANHOURS PER UNITS LISTED

Item	Unit	Manhours		
		Carpenter	Truck Driver	Total
Single Swing Door				
Frame or jambs	opening	1.40	.05	1.45
Hang door	opening	.90	.05	.95
Install trim	opening	1.40	.03	1.43
Double Doors				
Frame or jambs	opening	1.70	.07	1.77
Hang doors	pair	1.20	.07	1.27
Install trim	opening	1.90	.05	1.95
Transoms				
Frame or jamb	opening	.50	.02	.52
Install sash	opening	1.00	.03	1.03
Install trim	opening	.60	.03	.63

Manhours include handling, hauling from job storage and installing interior doors and transoms as outlined above.

Manhours do not include the placement of hardware or glass and glazing. See respective tables for these charges.

PREHUNG DOORS

MANHOURS REQUIRED EACH

Door Size Width X Height	Manhours		
	Carpenter	Truck Driver	Total
1'4"x6'8"	1.10	0.05	1.15
1'6"x6'8"	1.12	0.05	1.17
1'8"x6'8"	1.15	0.05	1.20
2'0"x6'8"	1.19	0.05	1.24
2'4"x6'8"	1.22	0.05	1.27
2'6"x6'8"	1.26	0.05	1.31
2'8"x6'8"	1.30	0.05	1.35
3'0"x6'8"	1.33	0.05	1.38
3'6"x6'8"	1.37	0.05	1.42
4'0"x6'8"	1.40	0.05	1.45

Manhours are for installation of factory assembled prehung interior flush hollow core door with tempered hardboard faces, solid fir jamb set and ½"x1⅝" casing, one pair of 3½"x3½" dull brass hinges, and brass pass lock.

Included in the time frames are handling, hauling from job storage, setting, plumbing, and complete installation.

FOLDING DOORS AND ROOM DIVIDERS

MANHOURS REQUIRED EACH

Door Size Width X Height	Manhours			
	Carpenter	Laborer	Truck Driver	Total
Wood Folding Doors and Room Dividers				
2'8"x6'8½"	1.30	0.50	0.20	2.00
3'0"x6'8½"	1.30	0.50	0.20	2.00
4'0"x6'8½"	1.30	0.50	0.20	2.00
5'0"x6'8½"	1.46	0.56	0.23	2.25
6'0"x6'8½"	1.63	0.63	0.24	2.50
7'0"x6'8½"	1.79	0.69	0.27	2.75
8'0"x6'8½"	1.95	0.75	0.30	3.00
9'0"x6'8½"	2.02	0.78	0.30	3.10
10'0"x6'8½"	2.08	0.80	0.32	3.20
11'0"x6'8½"	2.15	0.83	0.32	3.30
12'0"x6'8½"	2.21	0.85	0.34	3.40
13'0"x6'8½"	2.28	0.88	0.34	3.50
14'0"x6'8½"	2.31	0.89	0.35	3.55
15'0"x6'8½"	2.34	0.90	0.36	3.60
Vinyl Fabric Folding Doors				
2'8"x6'8"	1.56	0.60	0.24	2.40
3'0"x6'8"	1.56	0.60	0.24	2.40
4'0"x6'8"	1.56	0.60	0.24	2.40
5'0"x6'8"	1.72	0.66	0.27	2.65
6'0"x6'8"	1.90	0.73	0.30	2.93
8'0"x6'8"	2.24	0.86	0.35	3.45

Manhours include handling, hauling, and installing doors, track channel or track, track gate where required, nylon glides, matching valance, and privacy latch.

WOOD SASH & SCREENS

MANHOURS PER UNITS LISTED

Item	Unit	Manhours		
		Carpenter	Truck Driver	Total
Single Window				
Assemble frame	opening	.95	.02	.97
Set frame	opening	1.00	—	1.00
Fit & hang sash	opening	1.40	.02	1.42
Install trim	opening	1.20	.01	1.21
Double Windows				
Assemble frame	opening	1.60	.03	1.63
Set frame	opening	1.40	—	1.40
Fit & hang sash	opening	2.25	.03	2.28
Install trim	opening	1.90	.02	1.92
Triple Windows				
Assemble frame	opening	2.15	.04	2.19
Set frame	opening	1.90	—	1.90
Fit & hang sash	opening	3.25	.04	3.29
Install trim	opening	2.40	.03	2.43
Window Screens	each	1.10	.02	1.12

Manhours include handling, hauling from job storage and installing wood sash and screens as outlined above.

Manhours do not include the placement of locks and lifts or glass and glazing. See respective tables for these charges.

FINISH HARDWARE FOR WOOD DOORS & SASH

MANHOURS PER UNITS LISTED

Item	Unit	Carpenter Manhours
Full Mortised Butts	each	.15
Half Mortised Butts	each	.10
Mortised Lock	each	1.25
Cylinder Lock	each	2.20
Door Closer	each	.90
Door Check	each	1.00
Door Holder	each	.50
Push & Pull Bars	each	1.25
Kick Plate	each	1.25
Rim Night Latch	each	.25
Panic Bolts	each	3.20
Sash Lifts & Locks	set	.40

Manhours cover time as may be necessary to complete installation of above listed items including checking out of job storage and hauling to installation site.

Manhours do not include the placement of frames, doors or trim. See respective tables for these charges.

Section 13

METAL DOORS, SASH, GLASS & GLAZING

This section covers the installation of the various types of metal doors and sash and the glazing of metal doors and sash as might be encountered in an industrial plant.

The following manhour tables are average of several jobs and include all operations as may be necessary for the individual item listed in strict accordance with the notes as appear thereon.

For the installation of wood doors and sash, refer to Section 11, entitled "Carpentry & Millwork".

EXTERIOR HOLLOW METAL DOOR

MANHOURS PER UNITS LISTED

Item	Unit	Manhours		
		Carpenter	Truck Driver	Total
Single Swing				
Frame	each	3.40	.10	3.50
Door	each	3.20	.10	3.40
Trim	side	1.00	—	1.00
Transom	each	1.90	.05	1.95
Double Swing				
Frame	each	3.90	.15	4.05
Door	pair	6.10	.20	6.30
Trim	side	1.40	—	1.40
Transom	each	2.40	.05	2.45

Manhours include handling, hauling and installing of frames, doors, trim and transoms as are outlined above.

Manhours do not include placement of door hardware, glazing or painting. See respective tables for these charges.

EXTERIOR PRESSED STEEL DOORS

MANHOURS PER UNITS LISTED

Item	Unit	Manhours		
		Carpenter	Truck Driver	Total
Single Swing				
Frame	each	3.20	.10	3.30
Door & hardware	each	8.00	.20	8.20
Double Swing				
Frame	each	4.10	.15	4.25
Doors & hardware	pair	16.00	.40	16.40
Single Slide				
Frame, track & hangers	each	8.00	.10	8.10
Door & hardware	each	8.00	.20	8.20
Double Slide				
Frame, track & hangers	pair	16.00	.15	16.15
Doors & hardware	pair	16.00	.40	16.40

Manhours cover handling, hauling and installing of items as listed above including drilling and tapping of doors and frames for the receiving and placement of hardware.

Manhours do not include glazing or painting. See respective tables for these charges.

EXTERIOR ALUMINUM DOORS & FRAMES

MANHOURS PER UNITS LISTED

Item	Unit	Manhours		
		Carpenter	Truck Driver	Total
Single Swing				
Frame	each	3.80	.10	3.90
Door & hardware	each	8.70	.20	8.90
Transom	each	2.10	.05	2.15
Double Swing				
Frame	each	4.50	.20	4.70
Doors & hardware	pair	17.00	.50	17.50
Transom	each	2.90	.10	3.00

Manhours cover handling, hauling and installing of items as outlined above including drilling and tapping for hardware and the installation of hardware.

Manhours do not include glazing or painting. See respective tables for these charges.

INTERIOR HOLLOW METAL DOORS

MANHOURS PER UNITS LISTED

Item	Unit	Manhours		
		Carpenter	Truck Driver	Total
Single Swing				
Frame	each	3.10	.10	3.20
Door	each	2.90	.10	3.00
Trim	side	.90	—	.90
Transom	each	1.70	.05	1.75
Double Swing				
Frame	each	3.50	.15	3.65
Door	pair	5.50	.20	5.70
Trim	side	1.25	—	1.25
Borrowed Light (single size)				
Frame	each	3.10	.10	3.20
Trim	side	.90	—	.90
Transom	each	1.70	.05	1.75

Manhours include handling, hauling and installing of frames, doors, trims and transoms as outlined above.

Manhours do not include placement of door hardware, glazing or painting. See respective tables for these charges.

INTERIOR PRESSED STEEL DOORS

MANHOURS PER UNITS LISTED

Item	Unit	Manhours		
		Carpenter	Truck Driver	Total
Single Swing				
Frame	each	2.90	.10	3.00
Door & hardware	each	7.20	.20	7.40
Double Swing				
Frame	each	3.70	.15	3.85
Doors & hardware	pair	14.50	.40	14.90
Single Slide				
Frame, track & hangers	each	7.20	.10	7.30
Door & hardware	each	7.20	.20	7.40
Double Slide				
Frame, track & hangers	each	14.50	.15	14.65
Doors & hardware	pair	14.50	.40	14.90

Manhours cover handling, hauling and installing of items as listed above including drilling and tapping of doors and frames for the receiving and placement of hardware.

Manhours do not include glazing or painting. See respective tables for these charges.

INTERIOR ALUMINUM DOORS & FRAMES

MANHOURS PER UNITS LISTED

Item	Unit	Manhours		
		Carpenter	Truck Driver	Total
Single Swing				
Frame	each	3.60	.10	3.70
Door & hardware	each	8.25	.20	8.45
Transom	each	2.00	.05	2.05
Double Swing				
Frame	each	4.30	.20	4.50
Doors & hardware	pair	16.15	.50	16.65
Transom	each	2.75	.10	2.85

Manhours cover handling, hauling, and installing items as outlined above, including drilling and tapping for hardware and the installation of hardware.

Manhours do not include glazing or painting. See respective tables for these charges.

INTERIOR KALAMEIN OR METAL COVERED DOORS

MANHOURS PER UNITS LISTED

Item	Unit	Manhours		
		Carpenter	Truck Driver	Total
Lap Type (frame not included)				
Single swing	each	8.10	.10	8.20
Double swing	pair	16.20	.20	16.40
Single sliding	each	8.15	.10	8.25
Double sliding	pair	16.25	.20	16.45
Flush Type				
Single swing				
Frame	each	3.00	.10	3.10
Door	each	8.00	.20	8.20
Double swing				
Frame	each	4.00	.15	4.15
Doors	pair	15.90	.40	16.30

Manhours include handling, hauling and complete installation of items outlined above.

Manhours do not include painting. See respective table for this charge.

STEEL ACCESS DOORS

MANHOURS REQUIRED EACH

Door Size Width X Height	Manhours			
	Carpenter	Laborer	Truck Driver	Total
8"x8"	1.30	0.50	0.20	2.00
12"x12"	1.30	0.50	0.20	2.00
12"x18"	1.43	0.55	0.22	2.20
12"x24"	1.43	0.55	0.22	2.20
16"x16"	1.56	0.60	0.24	2.40
18"x18"	1.56	0.60	0.24	2.40
24"x24"	1.95	0.75	0.30	3.00
24"x36"	2.15	0.82	0.33	3.30
32"x32"	2.21	0.85	0.34	3.40
48"x48"	2.34	0.90	0.36	3.60

Manhours include handling, hauling from job storage, setting in position, aligning, and plumbing of access door and frame.

Manhours are average for installation in masonry, tile, sheet rock, or plaster walls.

ERECT STEEL & ALUMINUM SASH

MANHOURS PER UNITS LISTED

Item	Unit	Manhours		
		Iron Worker	Truck Driver	Total
Aluminum Sash				
Casement type	100 sq ft	11.70	.20	11.90
Fixed type	100 sq ft	9.90	.20	10.10
Steel Sash				
Projected type	100 sq ft	8.10	.20	8.30
Continuous monitor type	100 sq ft	9.00	.20	9.20
Vented casement type	100 sq ft	12.60	.20	12.80
Fixed casement type	100 sq ft	11.25	.20	11.45
Mullions	100 lin ft	12.00	.15	12.15
Sash Operators	100 lin ft	11.25	.25	11.50

Manhours include handling and installing of sash or sash items as outlined above.

Manhours do not include glass and glazing, painting or scaffolding. See respective tables for these charges.

GLASS & GLAZING SASH & DOORS

MANHOURS PER HUNDRED (100) SQUARE FEET

Item	Manhours		
	Glazier	Truck Driver	Total
Metal Wall Sash	6.00	.50	6.50
Monitor Sash	6.65	.50	7.15
Wood Sash	6.00	.50	6.50
Doors	6.10	.50	6.60
Thermopane (1")	28.80	4.00	32.50
Plate Glass Entrances	18.50	3.00	21.50
1/4" Plate Glass Store Fronts	17.00	2.85	19.85
Mirrors (average)	14.40	1.00	15.40
Clean Glass	4.50	—	4.50

Manhours include handling and placement of glass and the installation of putty or glazing compound and stops.

Manhours do not include installation of doors, sash or scaffolding. See respective tables for these charges.

CAULKING AND SEALANTS FOR DOORS AND SASH

HOURS PER HUNDRED (100) LINEAR FEET

Type and Size of Joint	Manhours		
	Carpenter	Laborer	Total
One Component Butyl			
¼"x½"	1.23	0.52	1.75
⅜"x½"	1.47	0.63	2.10
½"x½"	1.72	0.73	2.45
One Component Polyurethane Polymer			
¼"x¼"	1.47	0.63	2.10
⅜"x¼"	1.72	0.73	2.45
½"x¼"	1.96	0.84	2.80
¾"x⅜"	2.45	1.05	3.50
1"x½"	2.70	1.15	3.85
Two Component Polyurethane Polymer			
¼"x¼"	1.82	0.78	2.60
⅜"x¼"	2.10	0.90	3.00
½"x¼"	2.45	1.05	3.50
¾"x⅜"	3.08	1.32	4.40
1"x½"	3.36	1.44	4.80
Filler			
Polyurethane sponge	0.42	0.18	0.60
Polyethylene foam	0.42	0.18	0.60

Manhours include cleaning joints of grease, moisture, and other foreign matter and placing of caulking materials.

Joints deeper than ½-inch should be filled with polyurethane sponge or polyethylene foam.

Manhours do not include scaffolding. See respective table for this time frame.

SPECIAL FLOORING

Included in this section are many manhour tables covering the installation of special flooring and floor coverings.

The manhours include all labor as may be necessary for the described operation in accordance with the notes as appear thereon.

For installation of concrete or wood flooring refer to sections 8 and 11, entitled "Concrete" and "Carpentry & Millwork", respectively.

BRICK & MARBLE FLOORS & WALKS

MANHOURS PER UNITS LISTED

Item	Unit	Manhours			
		Mason	Marble Setter	Helper	Total
Brick Floors & Walks					
Acid proof floors	1000	12.50	—	10.00	22.50
Herringbone pattern	1000	36.00	—	22.00	58.00
Basket weave pattern	1000	30.00	—	20.00	40.00
Brick steps	1000	44.00	—	24.00	68.00
Marble Floor & Treads					
Marble floor slabs	100 sq ft	—	26.00	32.00	58.00
Floor tile					
Cement bed	100 sq ft	—	1.10	2.50	3.60
Tile	100 sq ft	—	12.50	12.50	25.00
Marble Base					
Smooth bottom plain	100 lin ft	—	14.00	18.00	32.00
Rough bottom plain	100 lin ft	—	20.00	25.00	45.00
Circular	100 lin ft	—	22.00	30.00	52.00
Rub Marble Floor	100 sq ft	—	6.00	—	6.00
Marble Stair Treads & Risers					
Treads	100 lin ft	—	17.50	19.50	37.00
Risers	100 lin ft	—	15.50	19.50	35.00
Marble Thresholds	each	—	.60	.60	1.20

Manhours include handling, hauling from installation stockpile and complete installation of items as outlined above.

Manhours do not include installation of sub-floors to receive above materials. See respective tables for these charges.

MARBLE WAINSCOT

MANHOURS PER UNITS LISTED

Item	Unit	Manhours			
		Marble Setter	Tile Setter	Helper	Total
Wainscot					
On level	100 sq ft	10.50	—	19.50	30.00
On rake	100 sq ft	18.00	—	27.00	45.00
Cap	100 lin ft	11.00	—	19.00	30.00
6" x 6" tile	100 sq ft	—	3.00	3.00	6.00
9" x 9" tile	100 sq ft	—	2.00	2.00	4.00
12" x 12" tile	100 sq ft	—	1.75	1.75	3.50
Base	100 lin ft	—	2.50	2.50	5.00
Clean & Wax	100 sq ft	—	—	2.50	2.50

Manhours include checking out of job storage or warehouse, hauling to site of erection and complete installation including all operations as may be necessary for the items outlined above.

Manhours do not include installation of sub-floors or floors to receive above items. See respective tables for these charges.

CERAMIC TILE

MANHOURS PER UNITS LISTED

Item	Unit	Manhours		
		Tile Setter	Helper	Total
Cement Fill	100 sq ft	1.10	2.50	3.60
Floor Tile				
3/4"-1" without border	100 sq ft	8.25	8.25	16.50
3/4"-1" with border	100 sq ft	9.75	9.75	19.50
2" hexagon	100 sq ft	13.00	13.00	26.00
3" hexagon	100 sq ft	15.75	15.75	31.50
2" or 3" square	100 sq ft	15.75	15.75	31.50
Base	100 lin ft	10.75	10.75	21.50

Above manhours cover checking out of job warehouse or storage facilities, hauling to site of erection and complete installation including all operations as may be necessary.

Manhours do not include placement of sub-floors placed prior to receiving above items. See respective tables for these charges.

TERRAZZO FLOORS

MANHOURS PER UNITS LISTED

Item	Unit	Manhours		
		Terrazzo Worker	Helper	Total
Sand & Felt Underbed	100 sq ft	1.75	1.75	3.50
Install & Finish				
Floors	100 sq ft	14.25	9.10	23.35
Border	100 sq ft	18.50	9.25	27.75
6" cove base	100 lin ft	21.25	21.25	42.50
Treads	lin ft	.40	—	.40
Wainscot	100 sq ft	20.00	15.00	35.00
Partition	100 sq ft	35.00	35.00	70.00

Manhours include handling, hauling and complete installation of items as are outlined above.

Manhours do not include the preparation of earth fill or placement of sub-floors prior to receiving above outlined items. See respective tables for these charges.

MASTIC FLOORS

MANHOURS PER HUNDRED (100) SQUARE FEET

Item	Unit	Manhours		
		Tile Setter	Helper	Total
Felt Underlay	100 sq ft	.30	.30	.60
Asphalt Tile Flooring				
6" x 6" tiles	100 sq ft	2.00	2.00	4.00
9" x 9" tiles	100 sq ft	1.75	1.75	3.50
12" x 12" tiles	100 sq ft	1.20	1.20	2.40
Cove base	100 lin ft	1.60	1.60	3.20
Rubber Tile Flooring				
4" x 4" tiles	100 sq ft	3.10	3.10	6.20
6" x 6" tiles	100 sq ft	2.25	2.25	4.50
9" x 9" tiles	100 sq ft	1.65	1.65	3.30
12" x 12" tiles	100 sq ft	1.15	1.15	2.30
Cove base	100 lin ft	1.65	1.65	3.30

Manhours include checking out of job warehouse or storage facilities, hauling to site of erection, placement of felt and/or cement, cutting of tiles or base, as may be necessary, and installation of tiles and base.

Manhours do not include placement of flooring prior to receiving above outlined items. See respective tables for these charges.

MASTIC FLOORS

MANHOURS PER UNITS LISTED

Item	Unit	Manhours			
		Tile Setter	Floor Mechanic	Helper	Total
Lino Tile					
Felt underlay	100 sq ft	.30	—	.30	.60
6" x 6" tiles	100 sq ft	3.50	—	5.50	9.00
9" x 9" tiles	100 sq ft	2.50	—	5.50	8.00
12" x 12" tiles	100 sq ft	1.75	—	5.50	7.25
Base	100 lin ft	2.75	—	2.75	5.50
Linoleum					
Laid direct on concrete	100 sq yd	—	11.70	7.20	18.90
Laid over felt	100 sq yd	—	14.40	10.80	25.20
Border	sq yd	—	.10	—	.10
Wax & polish	100 sq yd	—	4.05	4.05	8.10

Manhours include checking out of job warehouse or storage facilities, hauling to site of erection, placement of felt and/or cement, cutting of tiles, linoleum and base and complete installation.

Manhours do not include installation of flooring prior to receiving above outlined items. See respective tables for these charges.

QUARRY TILE & WOOD BLOCK FLOORING

MANHOURS PER UNITS LISTED

Item	Unit	Manhours				
		Tile Setter	Helper	Carpenter	Laborer	Total
Quarry Tile Flooring						
3" x 3" tiles	100 sq ft	17.25	17.25	–	–	34.50
4" x 4" tiles	100 sq ft	14.00	14.00	–	–	28.00
6" x 6" tiles	100 sq ft	13.50	13.50	–	–	27.00
9" x 9" tiles	100 sq ft	11.75	11.75	–	–	23.50
Base	100 lin ft	14.90	14.90	–	–	29.80
Wood Block Floors						
Over wood sub-floors	100 sq ft	–	–	3.75	1.25	5.00
In mastic over concrete	100 sq ft	–	–	4.00	1.60	5.60

Manhours cover handling, hauling and complete installation including all necessary operations of items as outlined above.

Manhours do not include installation of sub-flooring. See respective table for this charge.

Section 15

SPECIAL WALLS & CEILINGS

This section is included for the purpose of covering labor, in manhours, for the installation of such items as plaster and stucco and related items, and acoustical ceiling.

The manhours as listed are average, for the individual items, of many projects which were constructed under varied conditions in various parts of the country.

The tables include labor for all operations as may be necessary to install or construct the individual items, or block of work, in strict accordance with the notes as appear thereon.

FURRING & GROUNDS

MANHOURS PER HUNDRED (100) LINEAR FEET

Item	Manhours		
	Carpenter	Lather	Total
Wood Furring			
On wood	1.50	—	1.50
On masonry	3.10	—	3.10
Wood grounds	4.10	—	4.10
Metal Furring			
Furring only	—	2.35	2.35
Furring with ties & bars	—	4.10	4.10
Corner beads	—	4.10	4.10
Picture moulding	—	4.10	4.10
Cornerite	—	3.60	3.60

Manhours include handling and placing of furring and grounds as outlined above.

Manhours do not include framing or scaffolding. See respective tables for these charges.

LATHING

MANHOURS PER HUNDRED (100) SQUARE YARDS

Item	Manhours		
	Lather	Helper	Total
Wood Lath	8.00	1.00	9.00
Gypsum Lath	8.00	1.00	9.00
Insulating Lath	9.00	—	9.00
Metal Lath on Wood			
Studs	7.60	1.35	8.95
Furring walls	7.60	1.35	8.95
Partitions	7.60	1.35	8.95
Flat ceilings	8.00	1.35	9.35
Panel or arch ceilings	11.20	1.80	13.00
Beams & girders	17.60	2.70	20.30
Simple covers & cornices	15.40	2.70	18.10
Complex cornices	19.80	3.60	23.40

Manhours include all operations as may be necessary for the handling and installing of lathing as outlined above.

Manhours do not include wall framing or scaffolding. See respective tables for these charges.

METAL LATH

MANHOURS PER HUNDRED (100) SQUARE YARDS

Item	Manhours		
	Lather	Helper	Total
On Metal Furring			
Walls & partitions	8.80	1.35	10.15
Flat ceilings	10.80	1.80	12.60
Panel or arch ceilings	12.00	2.05	14.05
Beams & girders	17.80	2.70	20.50
Coves & cornices	16.00	2.70	18.70
Complex cornices	20.00	3.60	23.60
Suspended Ceiling			
Including hangers & channels	22.50	4.50	27.00
On rib lath including rods	18.00	4.00	22.00

Manhours cover handling and installing including operations as may be necessary for the complete installation of metal lath on the type construction as outlined above.

Manhours do not include installation of framing for walls or arches or the installation of beams or girders or scaffolding. See respective tables for these charges.

PLASTERING

MANHOURS PER UNITS LISTED

Item	Unit	Manhours		
		Plasterer	Helper	Total
Scratch Coat				
Regular surface	100 sq yd	6.00	6.00	12.00
Irregular or curved	100 sq yd	9.90	9.00	18.90
Brown Coat				
Regular surface	100 sq yd	7.75	7.25	15.00
Irregular or curved	100 sq yd	10.00	9.25	19.25
White Finish Coat				
Regular surface	100 sq yd	8.20	6.25	14.45
Irregular or curved	100 sq yd	12.80	8.00	20.80
On beams	100 sq yd	16.00	13.00	29.00
Ornamental	100 sq yd	24.00	17.00	41.00
Sand Finish Coat				
Regular surface	100 sq yd	9.80	6.50	16.30
Irregular or curved	100 sq yd	14.40	8.25	22.65
On beams	100 sq yd	16.80	14.00	30.80
Ornamental	100 sq yd	24.80	18.00	42.80
Keene's Cement	100 sq yd	18.50	7.50	26.00
Add for following if necessary.				
To simulate tiling	100 sq yd	12.80	1.50	14.30
Running arises	100 lin ft	6.40	—	6.40
Planting on mouldings	100 lin ft	27.20	—	27.20
Planting on bas-relief	100 lin ft	40.00	—	40.00

Manhours cover handling, mixing with automatic mixer and placing of plaster of the type and coat as listed above and include all operations as may be necessary.

Manhours do not include the installation of furring, grounds or lath or scaffolding. See respective tables for these charges.

STUCCO

MANHOURS PER HUNDRED (100) YARDS

Item	Manhours		
	Plasterer	Helper	Total
On Frame Construction			
Scratch coat	6.40	6.40	12.80
Brown coat	8.00	8.00	16.00
Plain float finish	12.00	9.90	21.90
Special Finishes			
Pebble dash	14.80	10.80	25.60
Broomed	14.40	10.35	24.75
Rough cast	16.00	9.00	25.00
Wash with Acid	5.85	5.85	11.70
On Masonry Construction			
Scratch coat	5.10	5.10	10.20

Manhours include handling, mixing with automatic mixer and placing of items as outlined above.

If special finishes above are used, substitute in place of plain float finish.

Manhours for brown and plain or special finish coats on masonry construction are same as those shown above for frame construction.

Manhours do not include the installation of furring, grounds, lathing or scaffolding. See respective tables for these charges.

WALL COVERINGS

MANHOURS PER UNITS LISTED

Item	Unit	Manhours		
		Painter	Helper	Total
Wall paper	C SF	0.90	0.90	1.80
Vinyl Wall Covering	C SF	0.61	0.61	1.22
Special Decorative Wall Covering	C SF	1.10	1.10	2.20
Flexible Wood Veneer on Fabric	C SF	2.75	2.75	5.50

All manhours include wall preparation. mixing and applying adhesive. cutting. placing. and finishing of material.

Wall Paper manhours are for installation of any quality of wall paper.

Vinyl Wall Covering manhours are for installation of polyvinyl chloride fused to a woven fabric.

Special Decorative Wall Covering manhours are for installation of glass cloth. felt or natural fabric. such as linen. cotton. or silk with paper or acrylic backing.

Flexible Wood Veneer on Fabric manhours are for installation of flexible wood veneers on fabric backing.

DRYWALL STEEL STUDS AND SUSPENDED CEILING SYSTEM

MANHOURS PER UNITS LISTED

Item	Unit	Manhours				
		Lather	Carpenter	Painter	Laborer	Total
Steel Stud System						
4" wall—studs 16" OC	C SY	13.97	–	–	6.83	20.8
4" wall—studs 24" OC	C SY	11.39	–	–	5.61	17.0
Suspended Ceiling System						
Channel 4'0" OC furring 16" OC	C SY	13.97	–	–	6.83	20.8
Channel 4'0" OC furring 24" OC	C SY	12.73	–	–	6.27	19.0
Drywall						
½" gypsum board—walls	C SF	–	1.34	–	0.66	2.00
½" gypsum board—ceilings	C SF	–	1.47	–	0.73	2.20
Tape joints—walls	C LF	–	–	3.65	–	3.65
Tape joints—ceiling	C LF	–	–	4.00	–	4.00

Steel Stud System manhours include placing metal runner tracks at floor and ceiling. placing 4-inch studs on centers as outlined, and placing one row of bridging.

Suspended Ceiling System manhours include placing 1½-inch ceiling channel and placing and tying ¾-inch by 2½-inch furring runner on centers as outlined above.

Drywall manhours include placing ½-inch gypsum board with screws and taping and finishing of joints as outlined above.

Manhours do not include painting or scaffolding. See respective tables for these time frames.

GUNITE & PARGETING

MANHOURS PER HUNDRED (100) SQUARE FEET

Item	Manhours				
	Plasterer	Helper	Iron Worker	Oper. Engr.	Total
Gunite Work					
Wire mesh	—	—	2.60	—	2.60
3/8" layer coat	.70	1.00	—	.35	2.05
Float or finish	1.05	.45	—	—	1.50
1/2" Pargeting					
Apply two coats	3.20	3.20	—	—	6.40
Clean & wash old surface	—	2.25	—	—	2.25

Manhours include handling and placing of items as outlined above.

Manhours do not include scaffolding. See respective tables for these charges.

ACOUSTICAL TILE CEILING

MANHOURS PER UNITS LISTED

Item	Unit	Manhours		
		Lather	Tile Setter	Total
Suspension System	100 sq ft	3.75	–	3.75
Metal Furring System	100 lin ft	1.90	–	1.90
Acoustical Tile				
Adhesive tape	100 sq ft	–	3.35	3.35
On wood furring	100 sq ft	–	4.00	4.00
Metal pan type	100 sq ft	–	4.25	4.25

Manhours include complete installation of items as outlined in above table.

Manhours do not include framing or scaffolding. See respective tables for these charges.

ROOFING & SIDING

It is the intent of this section to cover in man-hours the installation or application of various roofing materials, special roof decking, and special sidings for industrial buildings.

The following tables include manhours for operations as may be necessary for the installation of a particular item in strict accordance with the notes as appear thereon.

The manhours as listed are average of many projects of various size, nature and location throughout the country.

ASPHALT, TAR & GRAVEL ROOFING

MANHOURS PER SQUARE

Item	Manhours
	Roofers
Single Application	
Sheathing paper	.24
15 pound asphalt felt	.40
30 pound asphalt felt	.44
Coat-coal tar pitch	.26
Coat-roofing asphalt	.26
Roofing gravel	.80
Built-up Tar & Gravel Roofing	
Three-ply	2.41
Four-ply	3.11
Five-ply	3.81
Built-up Asphalt & Gravel Roofing	
Three-ply	2.54
Four-ply	3.34
Five-ply	4.54

Manhours include handling and placing of roofing and roofing items as outlined above, including installation of rope and pulley hoist to convey materials to roof.

Manhours do not include scaffolding. See respective table for this charge.

INSULATION BOARD & FLASHING

MANHOURS PER UNITS LISTED

Item	Unit	Manhours		
		Roofer	Helper	Total
Insulation Board				
1/2" thick	square	.60	—	.60
1" thick	square	1.30	—	1.30
1-1/2" thick	square	2.00	—	2.00
2" thick	square	2.55	—	2.55
Flashing				
Metal cap	100 lin ft	7.20	2.00	9.20
Metal trough	100 lin ft	6.40	2.00	8.40
Composition	100 lin ft	3.20	—	3.20
Counter Flashing				
Composition	100 lin ft	3.20	—	3.20
Metal	100 lin ft	7.20	2.00	9.20
Roof Waterproofing				
Metal valleys	100 lin ft	6.00	2.00	8.00
Metal ridge roll	100 lin ft	6.24	2.00	8.24
Metal gutter	100 lin ft	5.96	5.60	11.56
Metal drip cap	100 lin ft	7.20	2.00	9.20
Metal gravel stop	100 lin ft	6.96	2.50	9.46
Smooth composition roll roofing	square	1.00	—	1.00
Roll roofing mineral surface	square	1.50	—	1.50

Manhours include handling and placing of roof insulation and roofing items as outlined above.

Manhours do not include mop-on, graveling or scaffolding. See respective tables for these charges.

If craft other than that listed above is used for the installation of listed items, due to craft jurisdiction, substitute proper craft for above manhours.

ROOF PLANK & TILE

MANHOURS PER SQUARE

Item	Manhours			
	Mason	Helper	Hoist Engr.	Total
Gypsum Roof Plank	1.10	3.60	.03	4.73
Interlocked Roofing Tile	3.40	1.90	.03	5.33
Roofing Slate	3.00	3.00	.03	6.03
Concrete Roof Plank				
Lightweight	1.30	2.20	.03	3.53
Ordinary precast	1.80	3.60	.03	5.43

Manhours include handling and placing of roof planking and tile as outlined above.

Manhours do not include installation of built-up roofing or scaffolding. See respective tables for these charges.

ROOF HATCHES AND PLASTIC DOMED SKYLIGHTS

MANHOURS REQUIRED EACH

Item	Manhours			
	Carpenter	Laborer	Hoist Engineer	Total
Roof Hatches				
3'0"x2'6"	1.20	0.60	0.20	2.00
2'6"x4'6"	2.40	1.20	0.40	4.00
2'6"x8'0"	3.60	1.80	0.60	6.00
Plastic Domed Skylights				
14¼"x14¼" through 30¼"x30¼"	0.60	0.30	0.10	1.00
30¼"x46¼" through 55"x55"	0.90	0.45	0.15	1.50
57½"x69½" through 92½"x92½"	1.50	0.75	0.25	2.50

Manhours include handling and placing hatch or dome and factory furnished curb and hardware.

Manhours do not include placing roofing or other roofing items. See respective tables for these time frames.

If craft other than that listed is used, because of craft jurisdiction, substitute proper craft for above manhours.

CONCRETE ROOFING & ROOFING ITEMS

MANHOURS PER UNITS LISTED

Item	Unit	Manhours			
		Cement Finisher	Laborer	Hoist Engr.	Total
Roof Fill					
Ordinary concrete	cu yd	1.50	3.70	.04	5.24
Lightweight concrete	cu yd	1.60	4.80	.02	6.42
Roof Saddles					
Ordinary concrete	cu ft	.10	.10	.03	.23
Lightweight concrete	cu ft	.12	.12	.03	.27
Cant Strip					
Ordinary concrete	100 lin ft	3.40	4.50	.05	7.95
Lightweight concrete	100 lin ft	5.10	5.10	.05	10.25

Above manhours include installation of concrete roof fill, saddles and cant strip as outlined.

Manhours do not include placement of built-up roofing or scaffolding. See respective tables for these charges.

METAL & WOOD ROOFING

MANHOURS PER SQUARE

Item	Manhours				
	Sheet Metal Worker	Carpenter	Laborer	Hoist Engr.	Total
Metal Roof Deck	2.80	—	.75	.50	4.05
Flat Seam Metal Roofing	4.00	—	5.50	—	9.50
Wood Shingles	—	3.40	1.60	—	5.00

Manhours include handling and placing of roofing items as outlined above, including hoisting or placing of materials at correct elevation.

Manhours do not include installation of wood decking or scaffolding. See respective tables for these charges.

If craft other than that listed above is used for the installation of listed items, due to craft jurisdiction, substitute proper craft for above manhours.

METAL ROOFING, SIDING & PANELS

MANHOURS PER HUNDRED (100) SQUARE FEET

Item	Manhours			
	Carpenter	Helper	Truck Driver	Total
Corrugated Steel				
On wood frame	1.05	1.05	.10	2.20
On steel frame	1.10	2.75	.10	3.95
Corrugated Aluminum				
On wood frame	1.10	1.20	.10	2.40
On steel frame	1.20	2.90	.10	4.20
Stainless Steel				
On steel frame	3.00	4.00	.10	7.10
Metal Facing Panels				
Insulated wall panels	6.00	2.00	.15	8.15
Galbestos	4.00	4.00	.15	8.15
Porcelain enamel	7.00	7.00	.20	14.20

Manhours are average for handling, hauling and placing, including bolting of items listed above on the type of framing shown.

Manhours do not include caulking or scaffolding. See respective tables for these charges.

If craft other than that listed above is used for the installation of listed items, due to craft jurisdiction, substitute proper craft for above manhours.

ASBESTOS AND ASBESTOS PROTECTED METAL ROOFING & SIDING

MANHOURS PER HUNDRED (100) SQUARE FEET

Item	Manhours			
	Carpenter	Helper	Truck Driver	Total
Asbestos				
Corrugated on wood frame	1.20	3.60	.10	4.90
Corrugated on steel frame	1.60	4.80	.10	6.50
Shingles	2.80	2.00	.10	4.90
Asbestos Protected Metal				
Roofing	3.83	3.83	.10	7.76
Siding	3.20	3.20	.10	6.50

Manhours include handling, hauling, placing and bolting or nailing of items as listed above on the type framing shown.

Manhours do not include additional or excessive caulking as may be necessary for sash, doors, etc., or scaffolding. See respective tables for these charges.

If craft other than that listed above is used for the installation of listed items, due to craft jurisdiction, substitute proper craft for above manhours.

Section 17

ORNAMENTAL METAL & SPECIAL PARTITION

This section is included for the purpose of covering, in manhours, the installation of ornamental metal building fronts and trims and special partitions.

This type of work is a highly specialized business and is usually received on the job prefabricated.

The following manhours are based on the assumption of installing prefabricated units and are average of many projects of the same nature installed under varied conditions.

ENTRANCES, CANOPIES, BUILDING FRONTS & METAL CLAD PLYWOOD FACING

MANHOURS PER LINEAR FOOT

Item	Manhours
Standard Rolled or Light Extruded Metal Building Fronts	
Alumilited Aluminum or Stainless Steel	
Simple construction	1.25
Average construction	2.90
Difficult construction	4.60
Heavy Extruded Aluminum Metal Building Fronts	
Simple construction	2.20
Average construction	4.50
Difficult construction	6.75

METAL CLAD PLYWOOD FACING

MANHOURS PER HUNDRED (100) SQUARE FEET

Item	Carpenter Manhours
Stainless Steel	14.00
Aluminum	12.50

Simple construction is assumed to be straight section with single door at one side of bay or no door.

Average construction is assumed to be recessed section with double door, soffits, corner and dividing bars.

Difficult construction is assumed to be double recessed section with double doors requiring additional bars, covers, etc.

Manhours are average for the type of work and operations as outlined above and are for either carpenters or sheet metal workers as the case may be.

Manhours do not include any wood work installation such as grounds or blocking. See respective tables for these charges.

PREFABRICATED METAL, METAL TOILET & WIRE MESH PARTITIONS

MANHOURS PER UNITS LISTED

Item	Unit	Manhours			
		Carpenter	Laborer	Truck Driver	Total
Prefabricated Metal					
Door sash & movable partitions	lin ft	2.75	1.00	.05	3.80
Metal Toilet Partitions					
Walls, brackets & headrails	each	2.40	.20	.05	2.65
Stiles or pilaster	each	.40	–	.03	.43
Posts	each	.75	.10	.03	.88
Doors & hardware	each	.50	.10	.05	.65
Wire Mesh Partitions					
On wood framing	100 sq ft	1.15	.15	–	1.30
On steel framing	100 sq ft	1.25	.15	–	1.40

Manhours for prefabricated metal and metal toilet partitions include handling, hauling and complete installation of items as outlined above.

Manhours for wire mesh partitions include handling, hauling, fabricating and erecting of items as outlined.

Manhours do not include painting, glazing or scaffolding. See respective tables for these charges.

Section 18

PAINTING

Painting is usually considered a specialty item throughout the construction field and as such is usually subcontracted to a contractor who specializes in this trade.

The purpose of this section is to afford the general construction estimator a good sound means or basis of estimating painting labor for industrial construction should he desire to make his own estimate or a check estimate to check his subcontractor's price.

We do not attempt to outline the method or procedure that an estimator should use in taking off this type of work. The manhour tables that follow are based on labor estimates made in accordance with the use of standard measurements used throughout the painting and decorating trades. The estimator is cautioned to be sure that he understands the various methods of measurement and measurement allowances which are used in the taking off of specialty items such as doors, sash, etc.

EXTERIOR IRON & STEEL

MANHOURS PER UNITS LISTED

Item	Unit	Painter Manhours			
		Prime Coat	First Coat	Add. Coat	Total Coats
Structural Steel					
Sandblast, scale & derust	ton	—	—	1.00	1.00
Wire brush, scale & derust	ton	—	—	5.20	5.20
Shop coat	ton	.70	—	.70	.70
Brush field coat (erected)	100 sq ft	—	1.00	.67	1.67
Spray coat (erected)	100 sq ft	—	.22	.15	.37
Miscellaneous Iron	ton	—	1.60	1.40	3.00
Metal Surfaces	100 sq ft	—	.48	.40	.88
Metal Trim	100 sq ft	—	.88	.85	1.73
Metal Deck	100 sq ft	—	.52	.50	1.02
Steel Sash (both sides)	100 sq ft	—	1.78	1.75	3.53
Metal Doors	100 sq ft	—	.56	.53	1.09

Manhours include handling, stirring, mixing and placing of paint on items as outlined above.

Additional coat manhours are for one coat only. Add same number of manhours for each additional coat applied.

Manhours for sandblasting are those of air tool operator; for wire brushing — those of laborer.

Manhours do not include scaffolding. See respective table for this charge.

EXTERIOR MASONRY, CONCRETE & STUCCO

MANHOURS PER HUNDRED (100) SQUARE FEET

Item	Painter Manhours		
	First Coat	Second Coat	Total Coats
Masonry Walls (brush)	.72	.72	1.44
Concrete Walls (brush)	.76	.76	1.52
Stucco Surfaces (brush)	.80	.80	1.60
Brick, Tile & Cement Walls (spray)			
Oil paint	.27	.27	.54
Cement water paint	.23	.20	.43
Synthetic - resin bound exterior paint	.25	.20	.45
Concrete Floors	.28	.25	.53
Waterproofing			
Common or smooth brick	.92	.52	1.44
Concrete, concrete block	.92	.52	1.44
Fine texture cinder block	.92	.52	1.44
Rough surface cinder block	1.00	.60	1.60
Cement plaster	.92	.52	1.44
Stucco	1.00	.60	1.60

Manhours include necessary labor to complete operations as outlined above.

If third coat is required, add same manhours as under second coat column.

Manhours do not include scaffolding. See respective table for this charge.

BRUSH EXTERIOR WOOD ITEMS

MANHOURS PER HUNDRED (100) SQUARE FEET

Item	Painter Manhours			
	First Coat	Second Coat	Third Coat	Total Coats
Wood Siding	.50	.45	.55	1.50
Wood Trim	.68	.65	.70	2.03
Doors & Windows				
Oil paint	.53	.72	.55	1.80
Enamel	.53	.80	.73	2.06

Manhours include handling, stirring, mixing and brushing on of type of paint on items as listed above.

Should third coat be unnecessary, eliminate manhours for this operation.

Manhours do not include scaffolding. See respective table for this charge.

INTERIOR SPRAY WORK

MANHOURS PER HUNDRED (100) SQUARE FEET

Item	Painter Manhours		
	First Coat	Second Coat	Total Coats
Flat Paint - smooth plaster walls	.22	.25	.47
Flat Paint - sandfinish plaster walls	.27	.28	.55
Industrial Enamel - smooth finish plaster	.22	.23	.45
Spray Gun Stipple	–	–	.83
Aluminum Paint			
Smooth finish plaster	.22	.22	.44
Sand finish plaster	.22	.22	.44
Concrete surfaces	.22	.22	.44
Cender blocks	.22	.22	.44
Calcimine or Coldwater Paint			
Brick, plaster & wood surfaces	.12	.12	.24
Casein Paint			
Over painted surfaces	.22	.25	.47
Over smooth finish new plaster	.23	.22	.45
Over sand finish new plaster	.25	.25	.50
Over cement or concrete surfaces	.28	.28	.56

Manhours include handling, stirring, mixing, filling spray gun and applying of type of paint on surfaces as listed above.

Spray gun stipple manhours are for the applying of stipple on pre-sized or pre-painted walls.

Manhours do not include scaffolding. See respective table for this charge.

INTERIOR WOOD FLOORS

MANHOURS PER HUNDRED (100) SQUARE FEET

	Painter Manhours			
Item	First Coat	Second Coat	Third Coat	Total Coats
Paste Filler	.57	—	—	.57
Paint	.33	.38	—	.71
Penetrating Stain Wax-hardwood	.25	.22	—	.47
Shellac	.25	.22	.18	.65
Stainfill	.38	—	—	.38
1 Coat Shellac - 2 Coats Varnish	.25	.32	.32	.89
2 Coats Shellac - 1 Coat Wax & Polish	.25	.22	.50	.97
Clean, Touch-up & Varnish	.75	.33	—	1.08
Clean, Touch-up , Wax & Polish	.75	.50	—	1.25
Floor Seal - on Maple, Pine & Oak	.20	.17	—	.37
Buffing Floors	.25	—	—	.25
Waxing over 2 Coats of Seal & Polishing	.50	—	—	.50
Linoleum Waxing	.50	—	—	.50

Manhours for paints, stains, shellac and varnish include handling, shaking, stirring and brushing on of items as outlined above.

Manhours for waxing and polishing include necessary time for hand placing of wax and polishing with machine.

INTERIOR SHEETROCK, PLASTER & MASONRY WALLS & CEILINGS

MANHOURS PER UNITS LISTED

Item	Unit	Painter Manhours			
		Primer Coat	Second Coat	Third Coat	Total Coats
Sheet Rock					
Filling	100 sq ft	—	—	—	.50
Taping, flushing & sanding	100 lin ft	—	—	—	3.00
Spackle & sanding	100 lin ft	—	—	—	1.68
Flat Finish					
Smooth finish plaster	100 sq ft	.33	.45	.40	1.18
Sandfinish plaster	100 sq ft	.60	.50	.47	1.57
Industrial Enamel					
Smooth finish plaster	100 sq ft	.37	.43	.51	1.31
Walls & Ceilings					
1 coat silicated texture finish	100 sq ft	—	—	—	.55
Rubberized flat finish	100 sq ft	—	—	—	.40
Odorless enamel finish	100 sq ft	—	—	—	1.46
Damp or Dry Masonry Walls					
Enamel finish	100 sq ft	—	—	—	.45

Manhours include handling, shaking, mixing, stirring and brushing on or placing of items as outlined above.

If last coat of flat finish on smooth finish plaster is to be brush stippled, add .45 manhours per hundred square feet.

Manhours do not include scaffolding. See respective tables for these charges.

MISCELLANEOUS FINISHES FOR INTERIOR PLASTERED WALLS & CEILINGS

MANHOURS PER HUNDRED (100) SQUARE FEET

Item	Painter Manhours			
	First Coat	Second Coat	Third Coat	Total Coats
Plaster Construction				
Semi-gloss, average textured	.75	.65	.60	2.00
Glazing & mottling, over smooth finish	1.02	—	—	1.02
Glazing & mottling, over sand-finish	1.33	—	—	1.33
Glazing & highlighting, textured	.88	—	—	.88
Starch & brush stipple, over glazed surface	.65	—	—	.65
Flat varnish & brush stipple, glazed surface	.70	—	—	.70
Smooth Finish Plaster				
Wash, touch-up 2 coats of gloss	.60	.60	.57	1.77
Wash, touch-up, size & 2 coats flat*	.60	.17	.83	1.60
Washing Smooth Finish Plaster	.53	—	—	.53
Washing Sand Finish Plaster	.75	—	—	.75

Manhours include handling, shaking, stirring and brushing on of items as outlined above.

If second or third coat is to be eliminated, deduct above manhours for coat or coats not desired.

*Above manhours for wash, touch-up and two coats of flat on smooth finish plaster are shown as follows:

> First Coat   — Wash and Touch-up
> Second Coat — Size
> Third Coat   — First and Second coat of flat

Manhours do not include scaffolding. See respective table for this charge.

MISCELLANEOUS INTERIOR BRUSH PAINT FINISHES

MANHOURS PER HUNDRED (100) SQUARE FEET

Item	Painter Manhours		
	First Coat	Second Coat	Total Coats
Washing Starched Smooth Surfaces & Restarching	.53	.65	1.18
Washing Off Average Calcimined Surfaces	.67	—	.67
Calcimining - water size	.17	.33	.50
Calcimining - oil size	.37	.33	.70
Calcimining & Brush Stippling	.40	.37	.77
Casein Paint over Painted Surfaces	.37	—	.37
Casein Paint over Smooth Finish - new plaster	.37	.32	.69
Casein Paint over Rough Sandfinish Plaster	.38	—	.38
Casein Paint over Cement Blocks	.57	—	.57
Casein Paint over Acoustical Surfaces	.58	—	.58
Casein Paint over Cinder Blocks	.75	—	.75
Smooth to Medium Rough Texture			
Synthetic resin-bound paint	.45	.37	.82
Bonding & penetrating oil paint	.55	.50	1.05
Wood Veneer Finish			
Lacquer finish	.27	.27	.54
Penetrating wax or synthetic resin	.18	.10	.28

Above manhours include time necessary for all operations as may be involved to complete the above listed items.

Washing and starching: Above first coat represents washing time and second coat represents starching time.

Calcimining: Above first coat represents sizing time and second coat represents calcimining time.

Above manhours do not include scaffolding. See respective table for this charge.

BRUSH WORK – INTERIOR TRIM

MANHOURS PER UNITS LISTED

Item	Unit	Painter Manhours			
		First Coat	Second Coat	Third Coat	Total Coats
Enamel - picture moldings & other trim, up to 6" wide	100 lin ft	.30	.52	.58	1.40
Spackling or "swedish putty" over flat surfaces	100 sq ft	1.25	–	–	1.25
Glazing & wiping over enamel	100 sq ft	1.25	–	–	1.25
Flat varnishing & brush stippling over glazed trim	100 sq ft	.45	.33	–	.78
Washing, touch-up & 1 coat enamel	100 sq ft	.53	1.10	–	1.63
Wash, touch-up, 1 coat undercoater & 1 coat enamel	100 sq ft	.53	1.00	–	1.53
Washing enamel finish	100 sq ft	.87	–	–	.87
Stain, shellac & varnish	100 sq ft	.35	.35	.43	1.13
Stainfill, shellac, gloss varnish & flat varnish	100 sq ft	1.25	.35	1.00	2.60
Penetrating stainwax	100 sq ft	.50	.40	–	.90
Polish penetrating stainwax (2nd coat)	100 sq ft	.58	–	–	.58
Wash, touch-up & varnish	100 sq ft	.50	.73	–	1.23
Waxing & polishing	100 sq ft	.85	–	–	.85
Synthetic resin finish	100 sq ft	.18	.15	–	.33

Manhours include handling, stirring and mixing as may be required and placing of materials as outlined above.

Enamel — Picture Moldings and Other Trim: First coat manhours is that of placing prime coat.

Flat Varnishing and Brush Stippling over Glazed Trim: First coat manhours are for varnishing and second coat manhours are for stippling.

Wash, Touch-Up and 1 Coat Enamel: First coat manhours are for wash and touch-up and second coat manhours are for enamel.

Wash, Touch-Up, 1 Coat Undercoater and 1 Coat Enamel: First coat manhours are for wash and touch-up and second coat manhours are for undercoater and enamel.

Stain, Shellac and Varnish: First coat manhours are for stain, second coat manhours are for shellac and third coat manhours are for varnish.

Stainfill, Shellac, Gloss Varnish and Flat Varnish: First coat manhours are for stain and fill, second coat manhours are for shallac, and third coat manhours are for one coat varnish and one coat flat varnish.

Wash, Touch-Up and Varnish: First coat manhours are for wash and touch-up and second coat manhours are for varnish.

Manhours do not include scaffolding. See respective table for this charge.

BRUSH INTERIOR – METAL WORK

MANHOURS PER UNITS LISTED

Item	Unit	Painter Manhours		
		Prime Coat	Second Coat	Total Coats
Miscellaneous Iron	ton	1.68	2.12	3.80
Metal Surfaces	100 sq ft	.56	.64	1.20
Metal Trim	100 sq ft	.80	.88	1.68
Metal Doors	100 sq ft	.60	.88	1.48

Manhours include handling, stirring and mixing, and brushing on of red lead or similar type of prime coat and finish second coat.

If third coat is required, add manhours for second coat to above total.

Manhours do not include scaffolding. See respective table for this charge.

PATENT SCAFFOLDING

Erect and Dismantle

DIRECT LABOR — MANHOURS PER SECTION

Length	Manhours Per Section					
	1 or 2 Sections High			More than 2 Sections High		
	Erect	Dismantle	Total	Erect	Dismantle	Total
1 to 2 Sections Long	1.40	1.00	2.40	1.70	1.20	2.90
3 to 5 Sections Long	0.90	0.60	1.50	1.00	0.70	1.70
6 Sections & More Long	0.70	0.40	1.10	0.90	0.50	1.40

Manhours are for installation of patent tubular scaffolding with 2" plank topping.   Sections are 7' long by 5' wide by 5' high.

Manhours include transporting scaffolding and materials from storage, erection, leveling, and securing scaffolding, installation of 2" planking, dismantling of scaffolding and transporting scaffolding and materials back to storage.

Manhours are for patent type tubular scaffolding only and are not intended to suffice for homemade type scaffolding.

CONVERSION TABLE

Minutes to Decimal Hours

Minutes	Hours	Minutes	Hours
1	.017	31	.517
2	.034	32	.534
3	.050	33	.550
4	.067	34	.567
5	.084	35	.584
6	.100	36	.600
7	.117	37	.617
8	.135	38	.634
9	.150	39	.650
10	.167	40	.667
11	.184	41	.684
12	.200	42	.700
13	.217	43	.717
14	.232	44	.734
15	.250	45	.750
16	.267	46	.767
17	.284	47	.784
18	.300	48	.800
19	.317	49	.817
20	.334	50	.834
21	.350	51	.850
22	.368	52	.867
23	.384	53	.884
24	.400	54	.900
25	.417	55	.917
26	.434	56	.934
27	.450	57	.950
28	.467	58	.967
29	.484	59	.984
30	.500	60	1.000

Printed and bound by CPI Group (UK) Ltd, Croydon, CR0 4YY

03/10/2024

01040432-0009